CONFESSIONS
of a Time Traveler
The essays, articles, and artwork
of a deep time junkie

By R. Gary Raham

The essays, articles, short stories and book excerpts were first published as follows:
"The view from a throne in dinosaur country" *Fossil News*, March 1999;
In the *North Forty News*: "Longs Peak perspectives," July 2001; "The social conquest of
Earth," July 2012; "Making post modern decisions with prehistoric brains," Sept. 2010;
"Ghost predator legacies," Feb. 2006; "Wellington through the ages," Aug. 2005; "Pollen:
Not just something to sneeze at," March 2004; "Some dumb animals are pretty smart," Jan.
2007; "Permafrost: biological treasure chest & climate wildcard," Nov. 2012; "Jefferson's old
bones and the spirit of a new nation," Jan. 2001; "Arthur Lakes and the sea monsters of fossil
creek," Oct. 2007; "Lauren Eiseley: haunted by the ghosts of Lindenmeir," June 2009;
"Visit to a challenging future," June 2007;
In *Colorado Gardener* magazine: "Pleistocene ghosts in a field or supermarket near you,"
April 2006; "Why are plants green?" May 2008; "Thank a flower today...for everything,"
Spring 2010; "Plants: not green wimps, but world changers," Summer 2009; "Plants are
aware, but do they care?" Educational Issue 2013; "Lichens and the hidden architecture of
the living world," Summer 2010; "Lynn Margulis: champion of the microcosmos," Education-
al Issue 2012; "The garden microverse," April 2009; "Alive & aloft in the aeolian zone,"
April 2013; "The mysteries of metamorphosis," Harvest 2014;
In *Trilobite Tales* newsletter: "Beware young fossil hunters," Oct. 2012;
"Beatrix Potter: hare raising illustrator and part time paleontologist," Sept. 2014;
The Dinosaurs' Last Seashore, Biostration 2010; *The Deep Time Diaries*, Fulcrum 2000;
"Elvinon's Wish" in ConAdian Souvenir Book for the 1994 World Science Fiction Convention;
A Singular Prophecy, Biostration 2011

ISBN-13: 978-0-9904826-5-9
ISBN-10: 0990482650
Library of Congress Control Number: 2015931874

Penstemon Publications

Wellington, Colorado
www.penstemonpublications.com

Table of Contents

iv   R. GARY RAHAM

INTRODUCTION
a prologue of confessions

I've been confessing for quite a few years now, mostly in front of middle school students who want to learn more about dinosaurs, fossils, ancient times and the madcap world of writer-illustrators. I chose "Confessions of a Time Traveler" as the unifying theme for my talks. It seems only fitting that this collection of my essays, articles, fiction and book excerpts takes refuge under the same title— even though these pieces were mostly directed at an adult audience.

Everybody loves a good confession. Sharing our primate obsessions, mistakes, and indiscretions is a big part of what makes us human. So here's my first confession: I still haven't quite decided what I want to do when I grow up and I'm a grandfather. Talk about dithering. I tell myself that indecision about what delights to pursue during a lifetime demonstrates cosmopolitan tastes. I'm sticking with that interpretation.

Second confession: My lifelong passions include a fixation on visual images, a preoccupation with existential questions, a fascination with the wonders of nature, and a profound streak of rationalism that tends to make me a professional doubter. This particular blend nudged me into science teaching followed by a 27-year career as a graphic artist. In my spare time, I dabbled around as a science fiction writer and later wrote science articles and books aimed at educational markets. I didn't truly become a time traveler until a fascination with fossils led me to the delights of paleontology. Trying to interpret and recreate the past from the remnants of what life leaves behind is both a humbling and inspirational pastime.

Third confession: I subscribe to the view that creative fictions tell the entertaining lies that highlight basic truths. My ancestors, like yours, sat around fires telling exciting stories to make sure the kids looked both ways before crossing the savanna. They created virtual realities with cave paintings instead of pixels and asked all the Big Questions about existence. In the process, they saw remnants of the past and tried to explain them. They dreamed of futures and proceeded to make some of them come true.

This collection concentrates on essays and articles written for the *North Forty News, Colorado Gardener* magazine, and other regional publications over the past twenty years, but the reader will also discover excerpts from longer works—including a bit of science fiction. I often chuckled as I wrote these pieces and hope you do too. Enjoy the view from a mountaintop. Learn about the microbial conquest of Earth, and how flowering plants remade the world. Visit Pleistocene ghosts at large in the modern world and discover how to make postmodern decisions with prehistoric brains. After finishing the book, you will be prepared to tell your child why plants are green and how to find Great Blue Dinosaurs along the meandering shorelines of majestic desert canyons. And I hope you'll enjoy my closing letter to our successors (of whatever species) 50,000 years hence.

This leads to my fourth confession: I plan to perform brain surgery on you. No worries. It's a non-invasive process used by everyone with a story to tell or an image to create. I plan to enter your mind and twiddle a few neurons with words and images in a blatant attempt to allow you to see the intricate edifice of life on planet Earth through my eyes. If that seems scary, it is. But it's what human beings—the master storytellers of planet Earth—do best. They read and rewire minds. Sometimes, their fictions even morph into new realities.

Our imaginations have converted uppity primates with inflated brains into a world-changing force of nature. Hopefully, we can use even more of that imagination to recognize that our powers are only effective if we use them to maintain the intricate biosphere that has evolved through the vast corridors of deep time.

In the lead essay I feast on a banquet of creation while perched on the humble pink throne of a field expedition to Utah. Primates with delusions of grandeur can always benefit from a perspective that places them squarely in the middle of nature's grand cycle. Enjoy the spread. If some of it proves indigestible, no matter. Whatever you leave behind will be food for someone—or something. Nature has performed magic every day since life began at least 3.8 billion years ago and you and I are the amazing result.

I know. That's a whopping confession for a rationalist to make.

I call the oil painting on page iv "Anomaly." It was originally created to illustrate a time-traveling science fiction story (unpublished) in which the protagonist dies in the past, leaving his remains next to those of 500-million-year-old trilobites. Creationists would love to find an anomaly like that!

The dragonfly, for me, symbolizes an evolutionary endurance. Their insect order has been around for 300 million years and counting. The prospects for H. sapiens *may not be so robust.*

Part 1:
Confessions to awe, ignorance, & wonder

ESSAY
The View From a Throne
in Dinosaur Country

I confess I couldn't resist visiting a dinosaur dig that kept Stephen Spielberg from being a liar in the movie, Jurassic Park. The Western Interior Paleontological Society (WIPS) provided volunteers to help John Bird and the College of Eastern Utah excavate a site they called Yellow Cat Quarry. The field trip I joined took place after the discovery of Utahraptor bones (a raptor large enough to meet the cast requirements for the movie), but we did work to recover some scale-like plates (scutes) of an armored ankylosaur.

High winds plagued earlier expeditions. We experienced 100+ degree temperatures that began to melt the enthusiasm of even die-hard paleontologists. We worked early, took midday breaks, and then pounded more rocks for a while in the late afternoon. Evening discussions and a chance to see the pristine Utah sky through an excellent telescope inspired this diatribe about time, space, and our ephemeral biological existence.

This essay first appeared in Fossil News, Journal of Amateur Paleontology, *Vol. 5, No. 3, March 1999.*

Just after dawn, I plodded up the trail to the throne with my camping shovel. The shovel was unnecessary except for my peace of mind. I, mighty hunter of long-dead dinosaurs, had heard coyotes howl outside my tent, and the shovel was like the jawbone of an ass--defensive armament against any canine Philistines that might choose to block my path. I shouldn't have worried. Not even a lizard contested my ascent. And, more importantly, the stick, which determined possession of the throne, was in the prone, or "unoccupied," position next to a prominent rock outcrop alongside the trail.

The throne soon appeared before me, an unassuming, waist-high platform erected over a pit and crowned with a pink toilet seat. For modesty's sake, it was placed behind a large, sandstone boulder. Someone built it with one-size-fits-all in mind, making it a bit large for me and reminding me of my innocent youth in Ypsilanti, Michigan.

Ah, but the view was grand.

All but the most megalomaniac of kings would have killed a herd of fatted calves for the view: a marble-cake desert valley of brown-

ish-red sandstones and gray mudstones stained with green washes of rabbit brush, yucca, and Mormon tea. Spread across the western horizon, spires and arches carved out of Entrada sandstone glowed with warm, russet, beach-sand swirls placed there some 130 million years ago. These rocks formed a true Valley of Kings because they housed a necropolis of once mighty dinosaurs that I would soon explore along the margins of the mesa top a short walk away.

Dinosaurs occupied a metaphorical throne of dominance on earth for 160 million years or so. They weathered a couple of major extinction events during the Mesozoic--an entire geological era in which they were the prime players in Earth's ecology. They finally succumbed to a confluence of disasters that included a series of volcanic eruptions on the future Indian subcontinent, the retreat of a vast inland sea that split North America, and the arrival of one or several Manhattan-sized rocks whose impact and fallout severely compromised their well being. But their long tenure on the planet left many bones--and sometimes, appropriately it would seem as I sit in residence upon the throne, clumps of fossilized dung for primates like me to ponder.

Academic deliberations in the coolness of a desert morning prove much easier than digging on a narrow ledge in the 100-degree heat of a desert afternoon while keeping one eye on an overhang of mudstone split with a worrisome-looking series of erosive cracks. Nevertheless, our small party of five paleontological volunteers, guided by John Bird, excavation foreman for the College of Eastern Utah's Prehistoric Museum, did find dinosaur bones. They belonged to ankylosaurs--armored vegetarians with plates and spikes meant to deflect the appetites of predators like *Allosaurus* and *Utahraptor*.

Our dinosaur quarry was famous for the first discovery of *Utahraptor*, thus making

Stephen Spielberg, the creator of the movie, Jurassic Park, a prophet of sorts. Spielberg knew about velociraptors--vicious three-foot-long Mongolian predators with sickle-shaped leg claws admirably suited for evisceration. He wanted bigger stars to be the villains for his movie, however, and invented some. In 1989, Rob Gaston, a sculptor and part-time bone hunter, found their twenty-foot-long, real-life counterparts while motorcycling through the backcountry of Utah. The fact that they were early Cretaceous in age, rather than Jurassic was a minor flaw, since many other stars of the film, like *T. Rex*, were Cretaceous dwellers as well.

We would have loved discovering raptor claws and skulls, but those are usually reserved for those who spend weeks, months, and years in the field, donating their sweat and passion in full-time chunks. Ankylosaur parts were exciting enough, especially when it took hours to free a vertebra from a layer of limestone and hours more to gently flake away mudstone from a stack of once-armor plates that now threatened to split apart with any false hammer strike. And once the sun cleared the mesa top about 10 a.m., my attention became divided between intellectual musings and paying attention to physical discomforts that could only be relieved by drinking gallons of water and crawling with my favorite lizard into any spot of shade available.

For hours we acted as agents of erosion, throwing hundreds of pounds of rock over a cliff, and quickly gained an appreciation of Deep Time, a phrase invented by the writer, John McPhee, to describe the time it takes to sculpt a planet's geological features. How long does it take to be covered by mud, which is covered by sand and water and more mud hundreds of feet thick until everything is crushed and compressed to stone? In the case of these dinosaurs it took about 100 million years.

Since I've now lived over a half a century, I can begin to comprehend what a 100-year interval is like. I have an outside chance of seeing my hundredth birthday. Facing the turn of a new millennium, I can vaguely grasp that a thousand years can be spanned by ten very long lifetimes. A hundred such long lifetimes could take us back to the beginnings of civilization and agriculture. But you would have to multiply that number by ten again to get back to the beginnings of our species and you would have to multiply that number by a thousand to comprehend the age of an ankylosaur who died near an ancient river whose mountain source has since been ground to dust and rubble.

If I started now and just counted to 100 million, one second per number, I could finish in three years.

So in the mornings, I sit lightly on the throne, knowing it has been rather hastily nailed together and that other creatures passed their time--and whatever else--on Earth for far longer.

Some say the attraction of dinosaurs, especially for kids, is their magnificent size and power, but part of it--because of our self-centered human conceits--may be imagining exotic creatures who dominated an Earth where people had no role at all, and their ancestors were just good snack food. Remains of one of the brainiest dinosaurs--a small ancestor-eater called *Troodon*--have been found near my throne. If poor *Troodon* had had the time to experiment with more brainpower and if he'd survived the Big Disaster, I suspect I would enjoy having a beer with him and swapping philosophies on a warm, starry evening.

As it turned out, I enjoyed those very activities with my fellow quarrymen. One volunteer brought his own hand-made reflecting telescope and entertained us with views of all manner of heavenly bodies. Orange-banded Jupiter rose early in the east showing off four of

his brightest moons. Saturn followed with an impressive halo of rings I hadn't seen through a telescope since my college days. These planets are near neighbors. You can reach Jupiter by rocket in several years or fly by light beam at 186,000 miles per second and make the trip in three quarters of an hour. Extend your trip another forty-five minutes to reach Saturn.

By day, bedeviled by Deep Time. By night, bewildered by Deep Space.

The distances even to close stars are far more daunting. Mizar (and its fainter companion, Alcor) in the middle of the handle of the Big Dipper are 88 light years away. The image of them I enjoyed through the telescope started their journey when my father was five years old. If one of them explodes today my children (or perhaps Grandchildren) will find out about it.

And we did see the remains of exploded stars, too. The Crab Nebula in Taurus, looking like a fourth of July cloud of gunpowder when the rocket flash is over, once flared brighter than Venus in the night sky of 1054 A.D. Chinese astronomers, as well as native Americans, duly recorded the event in court documents and on cave walls. Of course, it really exploded 6000 years before, but the light of the blast didn't reach Earth until then. The veil nebula in Cygnus looks like a feathery, broken smoke ring about 70 light years in diameter. Astronomers think its mother star blew up 40,000 years ago.

It would be nice to have a convivial chat with some alien intelligence and ask her what she thought of all this. Some people may tell you they've had such discussions, but I'm not convinced. Scientists have searched off and on for years for intelligent radio signals from space and have found nothing but radio beacons from burned- out star cores. The odds are stacked high against hearing messages from Deep Space. It's not that there aren't plenty of stars with planets and probably plenty of stars with life and many with intelligent life as well. I suspect there are. A key element in the equation is how long might a technological society capable of--and interested in--sending signals survive. Timothy Ferris in *The Mind's Sky* estimated that if such civilizations endured ten million years there might be 4,000 thriving in our galaxy today. But if such civilizations only endured 10,000 years, there might be as few as four alive and transmitting.

Any aliens listening to us would first hear the news of Warren G. Harding's election to the presidency in 1920.

So, next morning after an evening of pondering, I sat taller on the throne, unchallenged king of all that I surveyed. But I was feeling a little lonely. I could read bones like tea leaves and make some pretty good guesstimates about life on Earth a hundred million years ago, but would never have a beer with an intelligent dinosaur. I could marvel at the last stars fading with the morning light, but be pretty sure, even if someone were looking back, they could never swim the gulf between us. With sufficient time and money, I suppose, I could possess and transform all that lies before me and turn it into Ypsilanti-like bathrooms where the walls were an unintimidating arm's length away. Perhaps that, like my camp shovel, would provide some small degree of security from howling infinities.

But that didn't seem like much fun.

Besides, it would be downright dangerous. I need the rest of the living world. I need to leave space for animals I can eat and others that will wag their tails in appreciation of my superior intelligence. I need to leave space for plants I can weave into Levis and hammer into condominiums, not to mention ones that give me air to breathe. I need the broad, unencumbered view to spark my pondering. And, for recycling purposes, I certainly need the efforts of the hard-working molds and bacteria that labor beneath the vulnerable but well-placed eminence on which I sit: a throne where the view is vast, but psychologically manageable, and my immediate duty, at least, is unambiguously clear.

Bottoms up! (Pen & ink)

ESSAY
Longs Peak Perspectives

I suppose climbing Long's Peak as we entered our fifties gave my wife and me confidence that we still had a few reserves in the tank and could do exciting things. We shared the trail with a caravan of people of all ages, each trying to put notches on their hiking sticks or check something off their "bucket list," although that term hadn't been coined yet. We enjoyed the somewhat anonymous camaraderie, the challenge of doing something physically hard, and the visual banquet of mountains and celestial wonders.

As a writer, I was pleased with the human-to-mountain and mayfly-to-fisherman metaphor I chose. Every creature lives to reproduce, with an individual's perspectives shaped by how long it takes to accomplish the deed against the backdrop of the rest of creation.

We climbed the peak in 2000, and I wrote the article for the North Forty News *in July 2001. I made a bold declaration at the end of this article that was, I confess, more for dramatic effect than a true declaration of intent. That's the charm of a good story. Writers craft them to make sense of a reality that is really quite messy, confusing, and filled with other challenges.*

At 3 a.m., the trail up the flanks of Longs Peak had shrunk from its familiar daytime meander among light-dappled pines to a shifting pattern of rocks, sand, and forest litter captured in the narrow ellipse of light from our headlamps. "Our first fourteener," my wife, Sharon, said. As an ex-PE teacher who had mastered kayaking, Tae Kwon Do, and several other sports, this was just a walk in the park--even if it was Rocky Mountain National Park.

"Another fine mess you've gotten me into," I smiled--proud writing-amoeba-turned-ath-lete--as we puffed our way carefully upward, now and then letting other hikers, many less gray around the edges than ourselves, pass us by. Each group greeted us warmly, then moved on, the rainbow colors of their sporty polymers fading into the darkness ahead. Though strangers all, we felt a kinship with these fellow hikers as we prepared to spend the next ten to fifteen hours on the same upward path. Why, you might reasonably ask, would so many people select such a task voluntarily? It seemed insufficient to tromp up a mountain just because it was there.

It had something to do with perspectives, I decided. High places can be good at giving you new ones.

Viewing from on high makes everything else look small: rivers become silver ribbons, cities become fireflies of light, political boundaries become abstractions that disappear altogether. Personal problems pale to insignificance. Sharon and I climbed, enjoying lacy streaks of northern lights--a new experience for us--and watched as Perseid meteors flashed themselves into extinction in the upper atmosphere as we passed the last gnarled Ponderosas guarding the trees' final frontier on the mountain.

Dr. Gregory Cajete, a Tewa Indian from the American Southwest, has observed that in his culture mountains represent yet another perspective: the long view of time. When an elder says, "pin peyah obe" (look to the mountain) he means think about generations yet to come. Mountains exist on a time scale that impacts the living environments around them for thousands--even millions--of years.

Understanding mountains presents humans with a problem not unlike that of a mayfly trying to understand the life of a fisherman.

A mayfly lives happily for several years beneath river rocks, until metamorphosing into a fragile, airborne adult insect with two or four delicate wings and three slender tail filaments. Since fish like to eat these adults, fishermen create look-alike ties as fish bait. Adult mayflies live only to reproduce. Some live for a few days, some for as little as two hours. If the typical fisherman lives for 657,450 hours (75 years), a mayfly's life represents .0003% that of a fisherman. What kind of perspective can a mayfly get of a fisherman in one brief encounter? The fisherman's life, in turn is .0003% of 25 million years. My perception of the constancy and durability of Long's Peak would be quite different if I could experience it over that magnificent span of years.

During the last 25 million years, primeval forests of redwoods and bald cypress have shrunk as the world has warmed and dried. Familiar oaks and maples have partly replaced them but, more importantly, grasses and weedy plants have grasped the opportunity to thrive on dry, windswept plains. Like green phoenixes, their shoots have rebounded from the grazing of humpless camels, hornless rhinos, and hoofless horses. The seemingly endless interactions of the grazers and the grazed ultimately produced not only humps, horns, and hoofs, but also a new living ecology: the prairie. The Rocky Mountains swelled upward from the Colorado Plateau and rivers cut through the rising rocks to form canyons, including the Grand One in Arizona. The climate has chilled some more, glaciers have come and gone and... Oh, yes...great apes flourished along with some naked apes that began to play with tools and build civilizations.

We apes on Long's Peak that morning chose not to be naked, although the wind was relatively calm. Sunrise bled above the horizon between Chasm Lake and the boulder field. "I think our pace is a little slow," Sharon said, "but we need to save something for the end."

"Look at the peak," I said, "it's like a mountain of gold."

At the boulder field, we could see a ragged, partial ring of rock called the Keyhole. Just to the left of the Keyhole, light winked at us off a plaque commemorating a victim of the mountain, Agnes Vaille, who successfully climbed the peak in the winter of 1925, but then died of exposure. "Remember John Wesley Powell," Sharon said. "If a one-armed guy can climb this hill, than so can we!"

At the boulder field the trail disappears. A hiker can pick any one of a dozen paths over the lichen-splotched boulders, using the Keyhole as their target destination. We climbed through the Keyhole and received a great prize: a view of several mountain lakes at the foot of a ring of snow-capped thirteeners. We snapped pictures and picked our way along the traverse to the trough, a rubble-strewn incline that led to the home stretch.

About this time, we began to meet people coming back. "I'm from Iowa, and I've seen enough," one person said rather cryptically. Another confessed that their fear of heights had got the better of them. We commiserated with them, but felt proud that we were still going forward.

The trough looked like just another rubble field, but it was steep and at high altitude. We kept moving, now and then dodging a slide of rocks created by someone above us. After an hour we passed a sign declaring we were at 13,972 feet--98% of the climb completed. But then flesh succumbed--at least mine did. My left thigh began to cramp, I felt a little nau-

seous, and a light rain was making the rocks slippery. I looked up at the steep hand-and-toe hold climb up the home stretch. "I don't think so," I said to Sharon, "but I'll wait if you want to go on." She thought about it, I think, but then sat down. "I'm tired, too," she said.

I was grateful.

We were disappointed, but not discouraged. Even if we were mayflies, we still had time to look the fisherman in the eye another time. And, although an epiphany didn't strike us like a lightning bolt, we did absorb some fraction of that perspective-changing quality of mountains that make them so attractive. We were exhausted, awed, thrilled, surprised, and humbled--all in the course of twelve and three-quarter hours.

We'll be back to climb another day.

View of Long's Peak in winter (above)

View of "The Keyhole" on the day of the climb (left)

ARTICLE
The Social Conquest and Occasional Decimation of Earth

Colorado's High Park fire in 2012 combined with a recent reading of E.O. Wilson's The Social Conquest of Earth *sparked this article dealing with eusociality—the ability of social animals to sometimes disregard personal health and survival for the good of the group. This skill has brought great success to its practitioners. Among vertebrates* Homo sapiens *reigns supreme. The roughly 14,000 species of ant and 2,800 species of termites dominate ecologies on the invertebrate side of things. E.O. Wilson estimates that humans and ants have roughly equal world biomass totals, perhaps on the order of 300 million tons each.*

Ants don't ponder their successes, of course, or worry about their impact on the world as a whole. Humans do—and they should. Human impacts have occurred at instantaneous speed compared to the slow evolution of insect societies, and Earth's rheostats for adjusting to change are in danger of faltering.

These reflections on Wilson's book appeared in the July 2012 issue of the North Forty News. *The artwork is a photomontage with a view of the High Park fire as seen from Wellington in the background. Wilson contemplates his favorite insect face to face.*

Disturb ants' nests and soldiers boil out, mandibles gnashing, ready to fight. Workers move larval grubs to deeper sanctuaries. The queen abides in her chamber, poised to generate more eggs to replace fallen comrades. Disturb human communities, as the raging High Park fire is doing in northern Colorado, and fire fighters swarm to action. Relief workers provide food and shelter. The wealth of individuals and a nation buys resources to protect lives and limit damage. According to E. O. Wilson, Nobel Prize winning scientist and author, both ants and humans are demonstrating the amazing power and adaptive success of **eusociality**—the altruistic behavior of multigenerational social groups to defend home and hearth. Wilson outlines his latest ideas in *The Social Conquest of Earth* (W.W. Norton, 2012).

For some people, being compared to an insect the size of a rice grain may not be appealing. Others may only attend to ants when they invade a picnic basket or kitchen. Ants, though small, wield large ecological clout. Wilson estimates the combined worldwide weight of ants and humans to be roughly equal. Other insects belong to the eusocial club, including termites and some species of wasps and bees. Among vertebrates—animals with backbones—only naked mole rats, a few bird species, and perhaps African wild dogs flirt with effective, multi-generational eusocial behavior. Humans stand out because they have combined their primate social skills with enormously enhanced brainpower. They pass on their skills not only through the relatively slow mutation of genes, but also through the rapid dissemination of **memes**: ideas transmitted through the cultural tools of art and language.

Genetic changes have molded social insects over hundreds of millions of years. In the process, ants, termites, and bees have flourished, but they have also solidly integrated themselves into the living world—the biosphere. Ants farm fungi and milk aphids for a sugar boost. They jealously protect food and shelter their preferred plants from other animals. Ants return the dead to the earth and succumb to predation themselves, entering food chains at several levels. Plants enlist ants, bees, and other insects to assist in their sex lives. The success of social insects is reflected in the success of the communities they inhabit. They demonstrate the fact that cooperation among different species is often as—or more important than—the competition between them.

We humans took a fast track to eusociality that, while highly successful in the short term, plagues us with conflicting loyalties. Wilson says, "The origin of modern humanity was a stroke of luck—good for our species for awhile, bad for the rest of life forever." Selfishness, cowardice, and hypocrisy serve individual survival, but honor, virtue, and duty serve our group survival. He claims these contrary

impulses will always be at war with each other. But he does hold out hope that we can overcome our flaws. He's encouraged by the ways we are becoming more interconnected as one global, rather than many tribal, communities.

The High Park fire threatens the home of a teacher friend. Her recent e-mails to 67 friends and family members apprising them of her situation demonstrates this hopeful trend. Offers of help poured in—from neighbors, from acquaintances and colleagues, and from individuals who have put time and sweat into helping her and her family build this home. She also received a lot of prayers and words of hope—from Christians, Buddhists, spiritualists of other persuasions and even atheists. And now that the fire has consumed over 80,000 acres, help from State and national political organizations continues to arrive.

Humans excel at providing help to others during crises. In the deep past, it was the tribe that rallied its chiefs and priests and the gods they nurtured to bring aid and comfort. Later, cities and nation states did the same, protecting their own, while often coming into conflict with other political and religious groups. Today, we are slowly forging world-wide person-to-person connections that are breaking divisive barriers: barriers erected by differences in politics, religion, sexuality, and race that have served us reasonably well in the past as ways to preserve regional social groups—but have NOT served well at all to integrate our species effectively into the global web of life.

We get into trouble when politics leads to dogmatism and inaction.

We get into trouble when religion leads to tribalism and irrationality.

We get into trouble when science creates power without responsibility.

We get into trouble when language and art confuse rather than inspire.

I truly appreciate Wilson's comment that "History makes no sense without prehistory, and prehistory makes no sense without biology." We too often forget that we are not the masters of the world we are its consciousness. Because of our social success we are like a hippo thrashing around in a rose garden—where HIPPO stands for Habitat destruction, Invasive species dispersal, Pollution, Population excesses, and Overharvesting of resources. If we can effectively rally our considerable smarts, Wilson sees a bright future. "If we save the living world," he says, "we automatically save the physical world, because in order to achieve the first we must also achieve the second."

Although Wilson is a rationalist, he admits to a certain amount of blind faith—perhaps a product of the eusocial talents he, like the rest of us, possesses. "Earth," he believes, "by the twenty-second century, can be turned, if we so wish, into a permanent paradise for human beings, or at least the strong beginnings of one." The tools we will need are an ethic of treating each other with simple decency (the Golden Rule), the unrelenting application of reason (the scientific demand for testable proof), and an acceptance of what we truly are: an intelligent, socially adept fragment of an old and continuously evolving biosphere.

Scratchboard created as book cover artwork for Dr. Janice Moore at Colorado State University in Fort Collins.

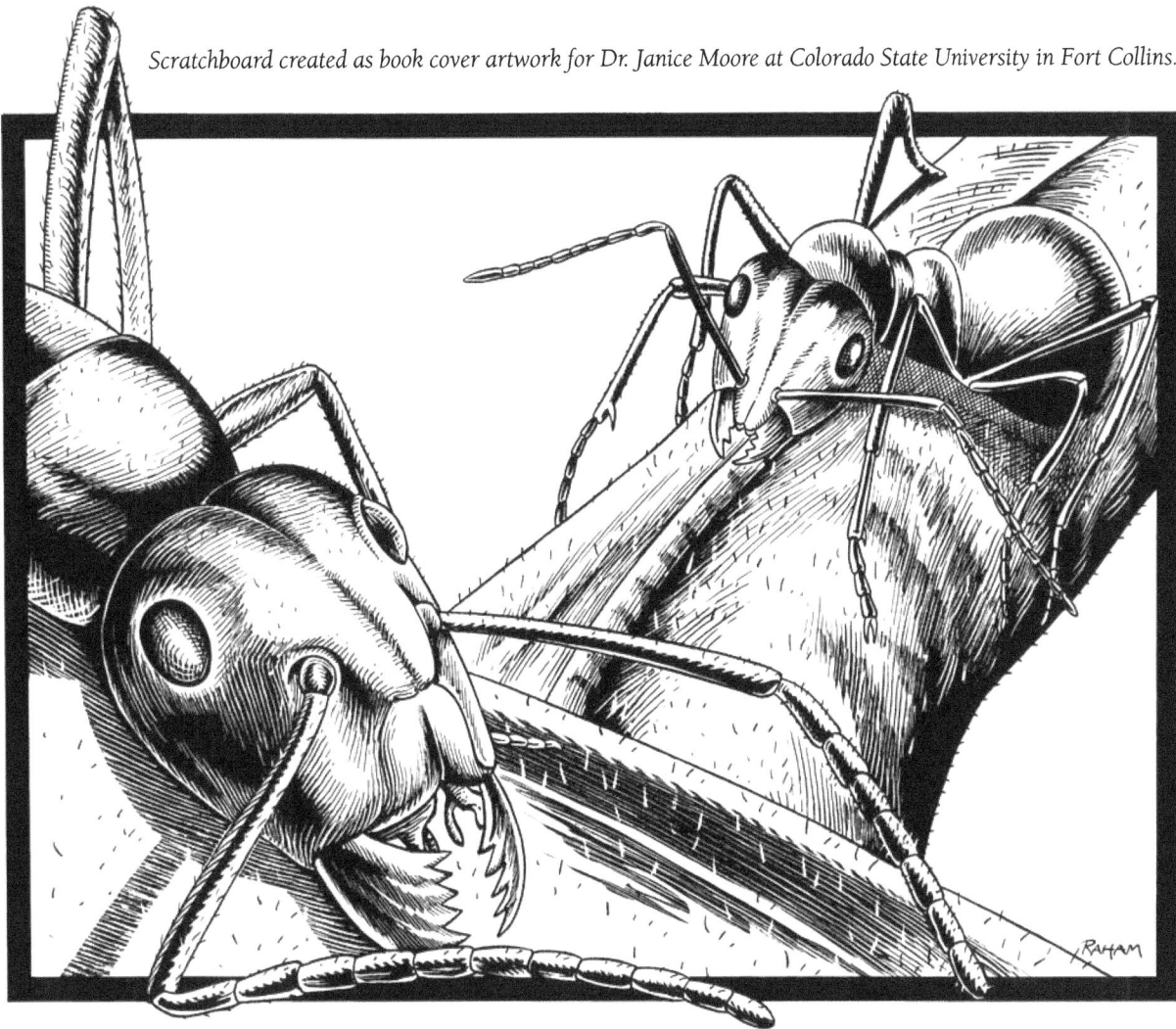

NEW MAN

RAHAM

ARTICLE
Making Post-modern Decisions
with Prehistoric Brains

Economic bad times tend to make us consider just why we make the choices we make—about what we buy, whom we decide to marry, whom we vote for in elections—and a whole host of decisions that may not always look so good in hindsight. I ran across an article by Dr. Ali Binizir in the Huffington Post *that addressed the question "Why do smart people make dumb decisions?" In the piece, he outlined some of what we have learned about behavioral economics. As naked apes whose cleverness has very quickly converted spears into intercontinental ballistic missiles and burning boughs into atom bombs, it behooves us to figure out just why we do what we do and when not to do it.*

When considering how to illustrate this piece, I immediately remembered the moronic features of Alfred E. Neuman from the old MAD *magazine of my mid-twentieth century youth. Bearded and dressed in the 28th century uniform of* Star Trek, the Next Generation, *he seemed a perfect embodiment for the discussion.*

This article appeared in the September 2010 issue of the North Forty News.

We've all said it: "How could I have been so dumb?" These days the particular act of dumbness may have involved stock market or mortgage decisions, but the complexity of our modern world—and the length of our modern lives—makes bad decision-making easy. In some ways, even though we may be among the smartest animals on the planet, we are under-qualified for picking the right food to eat, much less helping to run a global civilization. But knowledge is power. If we learn to recognize our human brain wiring biases we can compensate for them. A whole field of study has developed under the heading "behavioral economics."

As a species we have specialized in "smartness" almost to an extreme. We make tools, we Facebook ourselves around the world, we deduce and abstract and imagine...but the amazing, swollen "wetware" perched on our shoulders is much better adapted to avoiding a lion attack than picking a 401K. Even things as basic as mate selection and child rearing can go awry when we live to 80 instead of 30. Fifty percent of modern marriages end in divorce, and since kids are discouraged from leaving aged and infirm parents in caves, those parents have to plan for enough income to live in Sun City.

So what anachronistic biases do we carry from our prehistoric past and how do we surmount them? Some examples follow.

I'm smart, but you're just lucky.

We all cherish the illusion that we're a bit smarter than the other guy. Thus, I may get a fantastic writing gig because of my amazing wordsmithing abilities while Shakespeare was just lucky getting all those plays performed to rave reviews. Likewise, I lost some plum jobs because of the recession, but Steven King's sales are down because he just isn't very scary anymore. In behavioral-economics-speak, this is the "fundamental attribution error." Perhaps without the self-confidence to try and beat the odds we might all still be pounding rocks on the savannah, but this assumption can derail efforts to improve our skills.

Compensation Strategy: awareness of this common tendency is the best defense.

You see, I told you so.

The confirmation bias is our tendency to bolster existing opinions by gathering only information that supports those views. We tend to avoid or downplay contradictory information. Thus one rarely sees Republicans or Democrats at the other party's rallies (unless they want to sabotage them somehow) and Creationists rarely attend symposia on the evolution of cetaceans. Perhaps group solidarity had survival value, but it also leads to unexamined, and thus sometimes poor, choices.

Compensation Strategy: If you are really looking for the truth, start without preconceptions and explore both sides of the argument with equal diligence.

I'm not flying there. Didn't you hear about the crash last year?

The availability bias often pushes us to make decisions based on emotionally charged memories that may rarely reflect actual dangers. Seeing news images of hundreds of people dying in a spectacular plane crash can dull our intellectual awareness that cruising down the freeway every day can be a much more dangerous undertaking. When our forebears heard a roar behind them, it was prudent to expect it was probably a lion like last time and make enough noise to imitate a whole tribe of trouble for the cat, but today's dangers are often more subtle and complex.

Compensation Strategy: Slow down the decision-making process enough to collect all the available and relevant facts.

Yeah, well, that's the way it is. It always has been.

We love to see patterns and make predictions based on them. Michael J. Mauboussin, an investment strategist, wrote in the March/April issue of *The Futurist* about the failure of induction—predicting future events from what's happening in the present—in an article entitled "Smart People, Dumb Decisions." He talks about the turkey that thought room and board with Farmer Brown was a pretty cushy deal—right up until the end of November when things took a sharp turn for the worse. Such an event is called a "phase transition" and is common in lots of phenomena like water suddenly turning to ice and global climate temperatures spiking out of control.

Strategy: Analyze the system to see if it could contain a rapid phase transition. Prepare for the full range of possibilities. Resist the temptation to treat a complex system as simpler than it is.

Behavioral economics gone awry

George Loewenstein and Peter Ubel in "Economics Behaving Badly" (*New York Times*, July 14, 2010) also urge people (and governments) not to forget the principles of old fashioned economics, either. Governments, for example, can label foods with nutritional information to induce people to make healthier choices, but if healthier choices are more expensive than junk food, individuals may choose the latter. One could either tax unhealthy foods or reduce subsidies on corn to make the costs of some sugary foods rise, for example. Which technique would be less objectionable? The authors conclude that, "Behavioral economics should complement, not substitute for, more substantive economic interventions."

Nevertheless, we need to keep in touch with that ur-woman or ur-man within us. Only forearmed with knowledge can we boldly decide what no one has decided before.

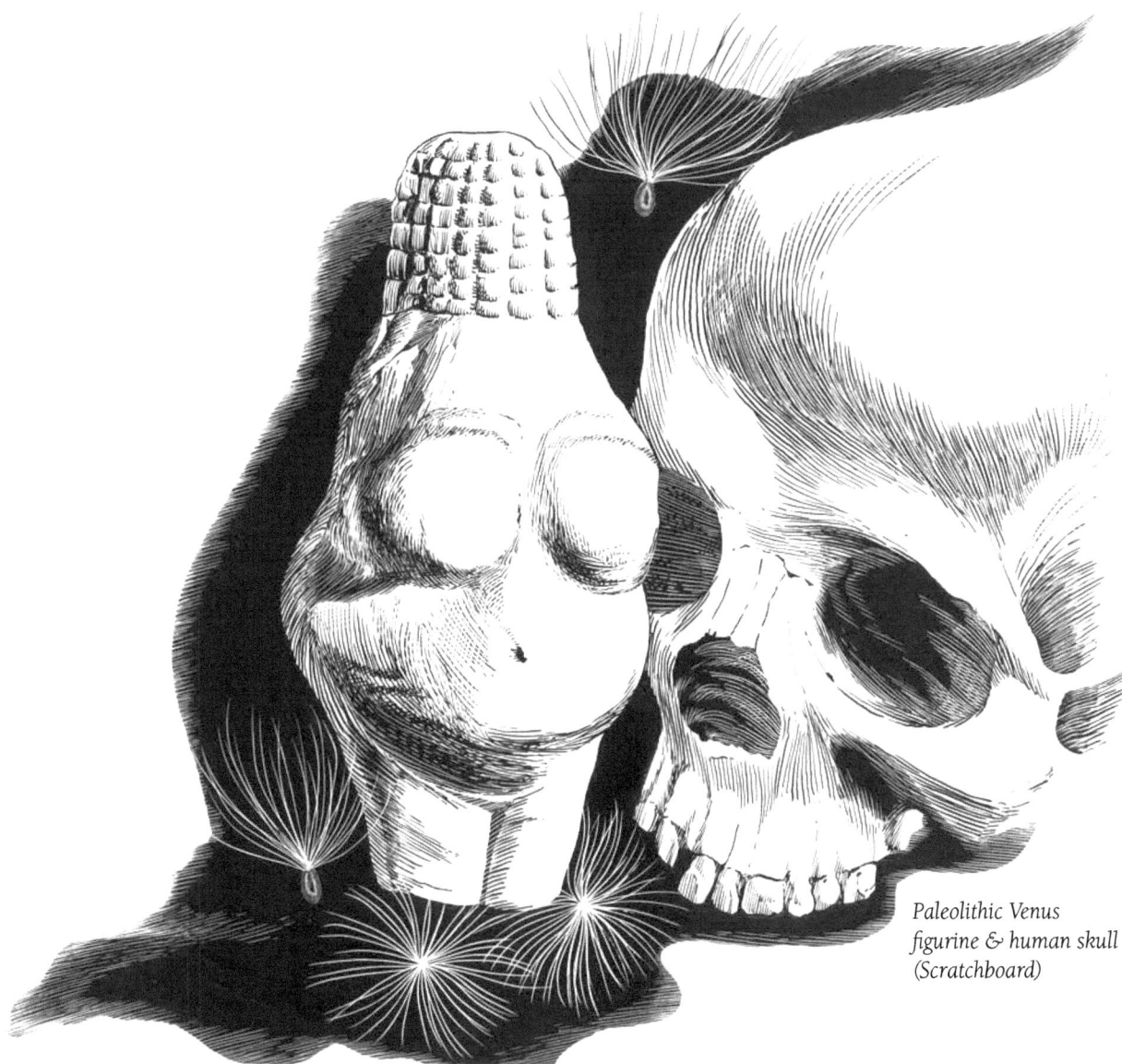

Paleolithic Venus figurine & human skull (Scratchboard)

Part II:
Confessing to amazing truths

ARTICLE
Ghost Predator Legacies

I worked with the Mike mentioned in this article when I was a graphic artist. He was a film stripper. I did my fair share of stripping, too—back in the day. It sounds exotic, but isn't. Strippers plied their craft by carefully positioning film negatives on plastic masking sheets so that they could be laid on light sensitive plates. Strippers aligned eight or sometimes 16 pages of a book, for example, on one large mask.

After cutting windows in the plastic around areas of the negatives that needed to be exposed, he or some other technician would "burn" the plates by exposing them to high intensity lights. After development with the proper chemicals, pressmen wrapped the plates around the cylinders of printing presses. The rotating plates, duly inked, transferred the positive images now inscribed on the plates to sheets of paper fed into the press.

Now, people sit at computer consoles and position images rather than handle film directly. Mike's job is essentially "extinct," like the ghost predators I talk about in this article. Mike also passed away at much too young an age some years after I left the printing company, but I remembered his hunting stories about sitting in blinds waiting for Pronghorns to amble by when I sat down to write this article.

Ghost predators lurk in our genomes. Lost friends and skills haunt our memories. I'm glad I can transfer a bit of what I know about both through the vehicle of this article first published in 2006 in The North Forty News.

A friend of mine named Mike hunted pronghorn. It really was more of an excuse to pass some time on the prairie admiring the big Wyoming sky, because "hunting" pronghorn involves creating a blind behind a decoy and waiting for a nosey animal to amble by. Mike ambushed pronghorn, because you aren't going to catch them easily any other way. Pronghorn canter at 25 miles per hour, gallop without straining at 45 miles per hour, and can run 56-62 miles per hour at top speed. They are the fastest animals in North America. No predator in their habitat today can catch an adult without ambushing them, unless they are sick or injured. The obvious question becomes why do pronghorn expend the energy for Ferrari-speed when Honda-speed economy would suffice?

John A. Byers, a pronghorn researcher, began asking himself this question in the 1980s when he studied the behavior of these superficially antelope-like grass-burners in Montana's National Bison Range wildlife refuge. The answer he deduced during 14 years of fieldwork involved ghosts—or at least ghostly survival imperatives etched into pronghorn genes by now extinct predators that could match their prey's speed.

Our prairies were once quite dangerous places—not a mere two hundred years ago, when wolves and bears called the Front Range home—but 10,000 years ago when post ice age hunters left spear tips in giant bison and mammoth vertebrae. Here's a list of the animal dangers that Pronghorn had to deal with just 5,000 generations in their past (Pronghorn females usually produce their first fawns at age 2):

• Dog-like predators included modern-style wolves, giant dire wolves (like those whose fossils have been recovered from California's LaBrea tar pits), wolf-sized and hyena-like plundering dogs (genus *Borophagus*), and two species of pack-hunting carnivores we have no modern name for (*Cuon alpinus* and *Protocyon*). Hyenas also prowled the prairies whose long limbs imply that they might have been fast runners.

• Bears: In addition to fairly modern types of black bears and grizzlies, the giant short-faced bear (*Arctodus simus*), 67% larger than a grizzly, possessed relatively long legs and feet oriented from front to back in a way implying that they were fast runners, too.

• Cats: Most people have seen pictures of the saber-toothed cat, *Smilodon*. This heavy-shouldered pussycat probably attacked big, slow prey like mammoth, cutting juggler veins with their impressive teeth. But fewer people know that our prairies were also home to a giant lion (*Panthera leo atrox*), a jaguar, and a cheetah, not to mention the familiar cougar (*Felis concolor*).

All these predators most likely gave mother pronghorns lots to worry about. Even though fawns can be up and running faster than a human in four days, they would certainly have been vulnerable to many of the predators listed above. Few of these predators, however, could have outrun a healthy adult. Wolves today, for example, rarely have been observed to take adult pronghorn, preferring instead to pick on bigger game, like elk. Based on the behavior of modern animals, it's also quite likely that the lions, jaguars, and cougars, would have ambushed pronghorn only when the right opportunity presented itself.

The behaviors of extinct predators, with no definitive modern counterparts with which to compare, are always somewhat problematical. We can speculate that the ice-age hyena and short-faced bear may have beset pronghorn, but they are different enough from modern species to raise questions about their dining tastes. Fossil cheetahs look enough like modern varieties to make more confident predictions. Modern cheetahs love to dine on African ungulates with pronghorn habits and size, and they can shift into running speed gears that can propel them at 60 to 70 miles per hour. Moreover, they coexisted with pronghorn for at least 2.5 million years—long enough, it would seem, to make speed a high priority for pronghorn survival.

Cheetahs chased pronghorns for 1.25 *million* generations. It's not surprising that pronghorn still run for their lives after just 5,000 generations of predator relief. 99.6% of their experience tells them to run and run fast. Various scientific studies have shown that behaviors in animals change slowly after selection pressure is relaxed—not surprising when such basic animal behavior gets hard-wired in the genes.

Other animals seem haunted by their pasts

as well. Hawks flying over large Madagascan lemurs cause fear reactions, even though none of these birds are large enough to harm them. Fossils show that large raptors once existed on the island that could have been a serious threat. California ground squirrel populations in California all show fear responses to snakes, although only some have snake predators in their modern habitats. Spanish horses that went wild in North America quickly grouped together in herds, following the instincts of forebears once threatened by many, if not all, the same predators as pronghorn.

The ice age still lives in all of us, which is why I can imagine Mike behind his blind, watching clouds build on the horizon, anticipating the nervous approach of a handsome buck, and wishing on some deep and visceral level that I were there, too.

The pronghorn, an ice age relic haunted by ancient predators

R. GARY RAHAM

ARTICLE
Pleistocene ghosts in a field
or supermarket near you

This piece nicely complements the previous article on ghost predators. While missing predators may haunt herbivores, the ghosts of seed dispersers often leave genetic relics in plant reproductive blueprints. Writers can't afford to waste research, so I created this article the same year (2006) for Colorado Gardener *magazine.*

Lindsay, my youngest daughter, served as the model for the grocery shopper depicted in the accompanying illustration. Illustrators can't afford to waste good models.

Ever felt haunted in the grocery store produce department? Perhaps felt the brush of shaggy fur on your arm near the papayas or heard muffled snorts while picking out an avocado for the chip dip? If not, such imaginings might just cross your mind after reading some of the theories of paleoecologists like Paul Martin, Dan Janzen, and Connie Barlow. The concept almost seems self-evident when introduced to it: many plants have evolved to attract seed dispersers that no longer exist. The evidence lives on in "over-built" fruit.

Our relatively short lives tend to skew our perceptions. Our parents or grandparents are the measure of old age, Thomas Jefferson is ancient history, and anything over 6,000 years old is, by many, unthinkable. But evidence of much older times lies all around. Thomas

Jefferson, in fact, became fascinated by mammoth and giant sloth fossils and helped demonstrate—through the explorations of Lewis and Clarke--that these huge beasts no longer existed, even in the wilds of North America. They were, in fact, extinct—a novel concept in the 18th century. Since Jefferson's time, paleontologists have found that North America was home to an entire zoo of megafauna (animals weighing more than a hundred pounds) that included primitive horses and camels, an American lion, saber-toothed cats, wolves, various members of the elephant clan (proboscideans), glyptodonts (think furry armadillo), toxodons (something like hippos), giant sloths, and more. These creatures all became extinct about 13,000 years ago at the end of the last ice age.

And they all had to eat something.

Granted, carnivores (for the most part) ate the veggie eaters, but the latter munched on plants. Very few flowering plants have become extinct since the glaciers melted away, although some have had their ranges severely restricted. And plants have a long history of developing various ways to attract animals to disperse their seeds. Plants like wild ginger attach lipid-rich food packets called elaiosomes to their hard-coated seeds. These attract ants, which eat the lipid gift and unintentionally bury the seed. Others surround their seeds with fairly small but pulpy, easy-to-pierce berries, which stand out from the green foliage in reds, blues, and purples. These are easy to spot and consume by sharp-eyed birds (and some mammals). The plant's seeds get physically or chemically roughed up while traversing the animal's gut, then are deposited at some distance from the parent plant to sprout. But there are other seed-bearing fruits that have no obvious means of dispersal—large, sometimes bounded by a hard rind, often with strong odors--these are the fruits that plants have created to woo large mammals.

Let's think about melons and gourds for a moment. In the American southwest, big-fruited gourds in the genus *Cucurbita* commonly grow along dry washes. These are commonly called skunk gourd because of the smell of their leaves. The fruit tastes bitter. Yet these viny plants are closely enough related to domestic summer squash that they could, and sometimes do, cross-pollinate with them. Typically, the large fruits rot in place. Rodents will sometimes bury gourds in their burrows, but only a fraction. They don't appear to be an effective disperser. But in Florida, mastodon fossils have been found in association with *Cucurbita* seeds. In Africa, the seeds of squashes in the same family can routinely be found in elephant dung. It doesn't seem like much of a leap to surmise that mastodons may well have been an important disperser for these vines.

In her book, *The Ghosts of Evolution*, Connie Barlow describes many other examples of North American plants with large fruits, often with smelly pulp, and seeds that need to be cut (scarified) or chemically treated to germinate properly, unless they have traversed a mammal's gut. Barlow was inspired by the research of botanist, Dan Janzen, and paleoecologist, Paul Martin, who teamed up to merge their scientific insights and propose a "megafaunal dispersal syndrome." The "wasted, rotting fruit" that Janzen found for many plant species in South America made sense when viewed in the light of the kinds of animals in their habitat, not only 13,000 years ago, but continually during the previous 30 million years or more. Their paper, entitled "Neotropical Anachronisms: The Fruits the Gomphotheres Ate," was published in the January 1982, edition of *Science*. It slowly, but steadily, has entered the consciousness of biologists, expanding the impact of ecological relationships through the deep time of geologists.

Gomphotheres, by the way, were oddly tusked elephants that, along with giant sloths that could rear up 18 feet, may have contentedly eaten papaya in Central and South America. The avocado you so casually use for chip dip comes from fruit originally designed to attract glyptodonts and toxodons. In North America, Osage orange trees produce huge fruits that deer and cattle can't handle, but horses love them. Horses were native to North America before they became extinct here at the end of the ice age. They were only reintroduced 400 years ago. Had they survived, Osage orange might not have become restricted, in the wild, to a few river valleys in eastern Texas. It grows fine in many upland habitats but can now only be dispersed by rolling or floating downhill.

Barlow said that once she became attuned to looking for vegetative anachronisms, every hike—even a walk in an urban park--turned into an adventure. The hard, coiled seedpods of Honey Locust trees conjure mastodons. Tall species of southwest prickly pear cactus (*Opuntia*) cry out for their ancient, ice age camel dispersers.

Even a routine trip to the grocery store can be an adventure filled with ghosts from deep time—once you know how to read the messages encoded in fruits made by plants that still remember the ice ages.

The following references may be of interest:

Barlow, Connie. "Ghost Stories From the Ice Ages," in *Natural History*, Volume 110, No. 7, September, 2001.

Barlow, Connie. *The Ghosts of Evolution, Nonsensical Fruit, Missing Partners, and Other Ecological Anachronisms*, New York: Basic Books, 2000.

Byers, John A. *American Pronghorn: Social Adaptations and the Ghosts of Predators Past.* Chicago and London: The University of Chicago Press, 1997.

Camelops, *the ice age North American camel (Pencil)*

ARTICLE
Wellington Through the Ages

Wellington, Colorado, my adopted town since 1977, celebrated its centennial in 2005. I was writing pieces for the North Forty News *at the time, so had the opportunity to write an article for the edition scheduled to be interred in a time capsule buried near the entrance arch from the old high school that became a decorative addition to the new middle school built that year.*

In 2001, a construction company's backhoe had turned up some mammoth tusks and teeth during construction of a new subdivision, so I decided to use readers' memories of that event as the starting point for a little more extensive history lesson.

I created the illustration shown on the opposite page for The Deep Time Diaries *(Fulcrum 2000), but it does depict an ice age scene perhaps not too unlike Wellington ten or fifteen thousand years ago.*

Happy 100th Birthday, Wellington! A hundred years is a long time but not too long. I can relate that venerable age to people and events important to me: my father was born in 1905, too. Longer time intervals get progressively harder to imagine, unless some physical reminder pops up, like the mammoth tusks and molars unearthed in a Wellington housing development four years ago, or the 10,000-year-old Lindenmeir folsom points and bison bones found just north of here in 1924.

What would this place we call home have looked like at various points in deep time –

times during which mountains rise and wash to rubble, and oceans come and go? The colorful, layer cake mountains and rolling plains we admire sometimes contain the clues that let geologists recreate these ancient landscapes. Some temporal stops might include the following:

10,000-20,000 B.P. (Before the present): Not a Safeway in sight. If you are a paleo-Indian at this time, best to be able to make or barter for the handsomely fashioned points which, when firmly attached to the haft of a spear, could bring down bison that stood 7 foot at the shoulder or other ice age fauna, like camels, giant sloths, and mammoths. Keep an eye out, of course, for saber-toothed cats, short-faced bears, giant wolves and other predators that might want to invite you for dinner. Gather edible plants watered by rivers and streams fed by melting mountain glaciers.

2 million years B.P.: Not a human in site, but the mountains would have looked much the same, although covered with considerably more snow during the height of glaciations. Ice reached depths of 1,500 feet in places. 5-7 million years B.P. a mountain growth spurt helped form favorite canyons like the Poudre and Big Thompson.

32 million years B.P.: Wellington would have been very hard to recognize. The modern Rockies, only half formed, poked through lush forest vegetation. Vast herds of vaguely pig-like grazing animals called oreodonts found the area inviting for millions of years, finally leaving the last of their bones in sediments near the small town of Grover northeast of us.

70 million years B.P.: The Rockies are just beginning their slow rise. Wellington is beachfront property for dinosaurs that can look east over the waves of a quiet inland sea that will ultimately leave shale deposits up to 14,000 feet deep to aggravate some modern gardeners. Today, the shells of clams turn up along road cuts, giant fish bones and scales get quarried along with limestone for cement, and now and then a dinosaur bone erodes free of sediment, like those of the "Masonville monster" (a relative of the meat-eating dinosaur, *Allosaurus*) found west of Horsetooth reservoir some years ago.

Wellington's history goes much deeper, but fortunately we don't have to dig for it. As the 1.8 billion-year-old granite that forms the core of the Rockies rose, it pushed up layers of sediments extending back to the days of a former mountain range (the ancestral Rockies) that existed here 300 million years ago. These layers have split and been pushed apart to stand as ragged ridges – the so-called "hogbacks" that stretch all along Colorado's front range. The oldest, sedimentary rocks lie nearest the mountain granite. The waters of Horsetooth Reservoir nestle among these hogbacks, sitting atop marine Paleozoic rocks 250 million years old. Rocks belonging to the "age of dinosaurs" rise to the east, culminating in a 100-million-year-old Dakota sandstone cap atop Reservoir Ridge, which is eroding back to beach sand.

Happy birthday, again, Wellington. You're not so old, after all, at least as measured on the geologist's deep time clock. May you build a rewarding future based on such solidly beautiful bedrock.

Inoceramus is a clam commonly found in local Cretaceous sediments. Oyster shells also erode from road cuts in nearby Weld County just east of town.

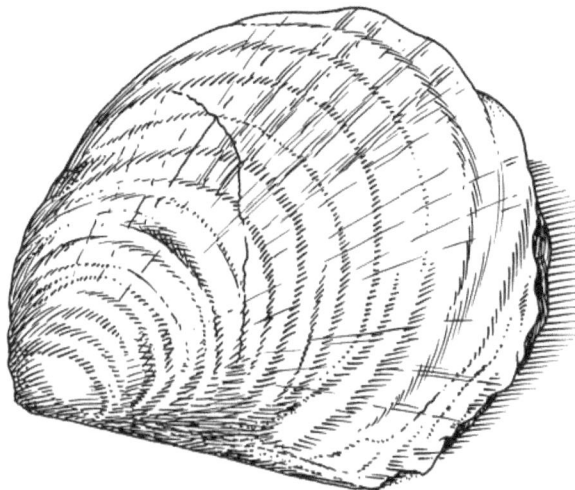

Wellington was beachfront property during the Cretaceous, as evidenced by shells and other marine life found in the area. This piece titled "Breakfast Patrol" originally appeared in The Deep Time Diaries.

I painted most of this illustration with acrylics used like watercolors, but scanned it into PhotoShop and made a few modifications to the sun and added some fog. It's always nice for artists to have options.

The planet depicted here evolved under the light of a redder sun, so plants use more of the available sunlight and reflect in more sombre tones of brown and black.

ARTICLE
Why are plants green?

Editors often suggest stories to writers. Jane Shellenberger with Colorado Gardener *magazine saw an article in* Scientific American *and thought it merited an interpretation for her readers. I'm glad she did. The idea appealed to me on several levels: as a fascinating aspect of biology, as a way to indulge my interest in life on other planets, as a way to reflect on how Earth's biotic colors are a function of the way life evolved and as an excuse to create a science fictional other-worldly illustration.*

Isaac Newton (1643-1727) used a prism to show that sunlight was composed of various frequencies (colors) of light. The German optician Joseph von Fraunhofer (1787-1826) demonstrated in 1814 that the spectrum was actually a spread of color punctuated with dark lines. We now know that the atoms of elements absorb some frequencies of light to create the black lines and emit other frequencies that our eyes recognize as colors. By 1860, the German scientists Gustav Kirchhoff (1824-1887) and Robert Bunsen (1811-1899) were able to show that each element absorbs and emits light in characteristic ways to produce a unique pattern— revealed in solar spectra. The spectrum from a star thus reveals the identity of its component elements as surely as a grocery store barcode.

This single discovery turned starlight into a Rosetta stone to interpret the chemical makeup of the entire universe…and to make reasonable speculations as to why plants are green. How amazing is that?

Comedian Bill Cosby in one of his old skits reminisced about dating a philosophy major. She asked bizarre questions like "Why is there air?" Some biologists, like Nancy Y. Kiang, who works for NASA's Goddard Institute for Space Studies, have been asking a question that might seem equally philosophical: Why are plants green? In Kiang's case, she hopes the answer to this question may help astronomers spot life on other planets from the relative comfort of their mountaintop or orbital observatories.

So, why are plants green and not blue, orange, or chartreuse? Are all plants green? Have they always been green? Might they be different colors on different planets? Colors are energy made manifest as light. Our perception of green foliage results from the mix of light photons that plants fail to capture with their chlorophyll pigments. Plants preferentially absorb high-energy blue photons and lower energy, but high volume red photons to build sugars from carbon dioxide and water during photosynthesis. Our cherished lawns and gardens reflect unabsorbed light in the green part of the spectrum back to our admiring eyes.

Because the sun produces all this dazzling light energy that runs the world, Kiang and others have scrutinized her stellar attributes. How does our sun compare with other stars that might be nurturing life-infested planets? Fortunately for us, our sun—designated as a yellow dwarf star—leads a long and boring (albeit incendiary) existence that has allowed life the time to evolve. It has been pouring out sunlight reliably for 4.6 billion years and can be expected to continue for another 5 or 6 billion years.

Bacteria first starting harvesting sunlight for energy 3.4 billion years ago. Bacterial pigments absorbed energy in the far-red and infrared (invisible to most animals) part of the spectrum and used sulfur and sulfate compounds to help build their cells. The reflected light from these deep-sea microorganisms would have looked distinctly purple.

By 2.7 billion years ago cyanobacteria—a.k.a. pond scum—began using chlorophylls and compounds called phycobilins to make sugars more or less like modern plants, churning out oxygen as a waste product that proved rather useful for animals sometime later. Cyanobacteria (blue-green algae in my old text books) cemented sand grains together to form huge high-rise pillars in the ocean called stromatolites. While stromatolites still grow in a few places, these primitive algae now usually form much more modest tangles of filaments or greenish-blue slime balls in suitably pond scummy conditions. But 2.7 billion years ago they made earth a blue-green planet. Her oceans didn't harbor green algae until about

750 million years ago and plants didn't venture onto dry land until about 475 *million* years ago.

You couldn't take green for granted until then.

Astronomers classify yellow dwarf stars like ours as "G" type stars. A star's letter designation signifies temperature in degrees Kelvin. The complete range of star types (from hottest to coolest) goes in the sequence O B A F G K M, which astronomers keep straight by remembering the phrase "Oh, Be A Fine Girl: Kiss Me!" (Those isolated observatories can be lonely.) Our G star falls toward the cooler end of the sequence. O, B, and A type stars tend to burn themselves out pretty quickly, but the other star types could burn long and reliably enough to nurture life on planetary systems. What kind of colors might we expect to see reflected from plants circling hotter or cooler stars than ours?

Hot F stars pour out light richer in the energetic blue photons. Land plants on planets circling such a star, assuming they had evolved a similar mechanism for aerobic photosynthesis, might need to reflect more blue photons to avoid damaging their tissues. Thus plants might display a distinctly blue tint. However, depending on the exact spectrum of light produced, the planet's distance from the sun, the photosynthetic pigments evolved, and the atmospheric composition, plants could reflect lots of red and yellow photons. Orange forests and lawns might be the norm for E. T.s living on planets in F star systems.

The majority of stars in our galaxy burn cooler than our sun and fall into the K and M categories. Planets in such solar systems would have to stick close to their parent star for adequate warmth and might be subject to more gravitational stresses because of that. Such cooler stars produce more low energy red light. To get enough photons to maintain photosynthesis, plants might have to absorb nearly the entire available spectrum. Plants might well appear clothed in somber shades of brown and black.

Plants and even microscopic photosynthesizers make their presence known at astronomical distances by more than the colors they reflect in visible light. Earth plants reflect heavily in the infrared, too. Scientists routinely map plant cover via satellite infrared imagery. Oxygen and ozone in the atmosphere are nearly sure signs of life. Free oxygen combines with iron ores and other minerals quickly unless replenished. In fact, banded iron formations on Earth are relics from the time when our planet first rusted during the early days of aerobic photosynthesis. Oxygen/methane combinations in an atmosphere and/or rising and falling concentrations of methane are hard to achieve except through metabolic activities. Nitrous oxide in the atmosphere results from plant decay. Abiotic sources like lightning produce only tiny amounts.

Cosby, an athlete as well as a comedian, told his girl friend there was no question why there was air: to blow up basketballs and volleyballs, of course. Kiang tells us plants on Earth are green because they are using as much of the rest of the sun's spectrum as they can to live long and prosper. While details of the photosynthetic story on other planets in the universe might vary, life processes will always give themselves away. A planet's living history glitters in the light it chooses to reflect.

Nancy Y. Kiang's article, "The Color of Plants on Other Worlds," appeared in the April, 2008, issue of *Scientific American.*

Varieties of microscopic life in a drop of water here on Earth

"Vorticella Forest" (above) Mixed media on acetate
"Golden Armour" (right) Mixed media

ARTICLE
Thank a flower today...for everything!

The writer Loren Eiseley (1907-1977) inspired this article. His elegant prose enchanted me when I first discovered his work in the 1970s, and his dark reflections often matched my own to the point where I adopted him as an honorary brother— even though he belonged to my parents' generation. When I moved to Colorado, I discovered he had been instrumental in an archeological discovery only 25 miles from my home in Wellington—the subject of the article on page 107.

Eiseley has a talent for making what could be mundane magical. He took something as commonplace as a flower and demonstrated how it became a world-changing opportunist. He exemplifies the power of the word to excite, to inform, and to serve as an engine of creative exploration.

During my research, I discovered that the common avocado produces a rather primitive style flower that provides a visual clue to what "Ur-flowers" may have looked like. They allowed me to illustrate how primitive flowers "exploded" after the End Cretaceous extinction event to transform our modern world.

This article appeared in the 2010 Spring issue of Colorado Gardener *magazine.*

Imagine gardens or mountain trails without the rainbow spatter of petals. Imagine open swells of land with carpets of fern instead of grass. Imagine somber green pines where aspen leaves flicker in the wind. This was the world for most of its existence— before flowers. In 1946, the writer/naturalist Loren Eiseley penned an essay entitled "How flowers changed the world." He acknowledged, as Charles Darwin had before him, that flowers evolved so quickly and left so little fossil evidence behind that their origins constituted an "abominable mystery." Today, paleontologists have uncovered more fossils and geneticists report that certain modern plants—one whose green fruit may have covered your last potato chip—provide clues to the family history of this revolutionary class of plants: the angiosperms.

Revolutionary? Come on. But it's true. Look outside and 89% of the plant life you see produces seeds nestled in the ovaries of flowers. Some flowers, like those of grasses, are not showy. They bloom modestly and toss pollen into the air, depending on wind to carry it to the appropriate female plant. Showy flowers advertise nectar with color and scent and induce insects, bats, or other creatures to visit them and, inadvertently, haul pollen across the meadow to complete their floral sex lives. Tasty fruits invite animal consumption and spread seeds to distant and perhaps moist, buried places.

Over 130 million years ago, the world's biosystems moved at a more ponderous pace. Evergreen (gymnosperm) forests, like the massive redwoods of California towered over animal giants, the dinosaurs. Plants, for the most part, reproduced with spores or naked seeds borne beneath the stiff bracts of cones. Their reproductive futures depended on favorable winds and climate. Plants had enlisted few animal propagation partners into the obligate arrangements flowers and insects now enjoy. Gymnosperms don't wrap their seeds in nutritious mounds of fruity tissue. The bracts that hold their seeds don't advertise with color and scent. In fact, "advertising" only makes seeds more vulnerable to predation and yields limited benefits.

At some point that changed.

Paleontologists like Dr. Nan Arens point to fossil discoveries that indicate flowering plants first exploited the dappled understories of Early Cretaceous evergreen forests, and over some tens of millions of years developed the metabolic pathways that allowed them to conquer "damp, dark, and disturbed" habitats. Then, fossils turn up in riverbank and other, more consistently sunny ecosystems. At this point, within a period as short as five million years, flowering plants dominated species counts in most floras instead of serving as dinosaur door mats. The climate changes (and asteroid impact) that ended the reign of dinosaurs also caused ecosystem disruptions in plant communities that may have given flowering plants their opportunity to profoundly

dominate world floras. Angiosperm talents for growing rapidly, producing seeds with built in energy sources, and enlisting animal partners proved invaluable during the ultimate "disturbed habitat" of Earth's last major extinction event 65 million years ago.

Pam Soltis, curator of molecular systematics and evolutionary genetics at the University of Florida Museum of Natural History, conducted a study that was reported May 2009 in *ScienceDaily* that provides hints about how the flower evolution revolution occurred at the molecular level. She studied the genome of the avocado, *Persea Americana*, which appears to possess very basic genetic instructions resembling metabolic pathways in certain cone-bearing plants. She considers the avocado—an example of a primitive, or basal angiosperm—to be a true genetic fossil. Her studies imply that the bracts of cones may, over time, have transformed into the (usually) green sepals, (usually) colorful petals, stamens (male organs), and carpels (female organs) of today's flowers.

Different studies show that other creative "mistakes" in early angiosperm development led to the endosperm tissue that forms the nutritious part of seeds—the parts we like to eat, chewing on corn or spooning up cereal—and the part that plant embryos use to fund their early growth. Dr. Paula Rudall and colleagues, working with *Trithuria*, a genus of small aquatic herb, has studied the early embryonic development of both the plant embryo and its endosperm. Based on those studies, she believes that endosperm originated as a monstrous embryo-in-waiting that failed to develop as a separate plant but did contribute to the resources of the surviving embryo.

So, flowers represent the radical offspring of staid evergreens, preening for animal helpers and feeding them high-energy snacks. Only incidentally do they entertain gardeners and curious hikers. They make us work for their favors, but give adequate, even extravagant, rewards. Eiseley concluded his essay by saying "The weight of a petal has changed the face of the world and made it ours," but perhaps that's too anthropocentric a viewpoint. Flowers changed the world to their own benefit and we came along for the ride.

Shucks. Go ahead and thank them anyway.

Like avocadoes, magnolias are considered a primitive family of flowering plants. Some very magnolia-like flowers may have adorned the habitat of dinosaurs like this Parasaurolophus. *This illustration is one of many in a graphic short story I published entitled* The Dinosaurs' Last Seashore *(see page 117)—a tale also inspired by the same Loren Eiseley quote.*

Yes, there's an animal in the middle of this picture—a tiger salamander to be exact—but the primitive plants around her helped make the invasion of land possible. Check out the article to find out how!

The artwork was painted on translucent acetate—the foreground details on the front, and the distant foliage on the back. The clouds were painted on blue matteboard and laminated together with the acetate.

ARTICLE
Plants: Not green wimps but world changers

Since I've already encouraged you to thank flowering plants for our modern world, you might as well thank the rest of the green world for making bare land a fit place to live. In this article, published in Colorado Gardener magazine in 2009, I outline how plants paved the way for the vertebrate invasion of land over 400 million years ago.

The illustration was created some time earlier in celebration of Colorado's only native salamander, the tiger salamander (Amblystoma tigrinum). For many years when my girls were growing up, we harbored a pet salamander (Sally) that we caught as an aquatic larva and watched it metamorphose into its adult, air-breathing stage—a developmental process that may mirror the evolutionary transformation of aquatic to terrestrial tetrapods.

Readers of *Colorado Gardener* understand the beauty and importance of plants. For many others, plants serve merely as salad fixins and pleasant backdrops for golf tournaments and Tarzan movies. Climatologists, paleontologists, and geologists, however, have recently made discoveries and developed models that reveal plants to be dominant players in the history of life and devoted caretakers of Earth's climate. If you haven't already, now is a good time to thank a green plant for its diligent efforts in creating the world we now enjoy.

Take a deep breath. The 21% of the air that's oxygen burns the food we eat and keeps us all energized and on the move. The air King Tut breathed also contained 21% oxygen, as did the air breathed by our earliest human ancestors. But such has not always been the case.

Rock formations formed before 2.5 billion years ago—particularly certain banded iron formations—demonstrate that Earth's atmosphere once contained virtually no oxygen. If it had, oxygen would have combined with exposed minerals to create the "red bed" formations seen in all subsequent millennia. What caused the rise in oxygen? Pond scum—mostly critters called cyanobacteria that practice the same trick as green plants. They use chlorophyll and other pigments to transform carbon dioxide, using energy from the sun, into sugars—leaving oxygen as a metabolic by-product. They started the job 3.5 billion years ago, saturated the ocean for a billion years or so, and then oxygen began building up in the atmosphere.

High oxygen levels in the oceans fueled first the development of oxygen burning microbes, then the symbiotic unions of those microbes into complex, single-celled creatures, and finally into the squirming, crawling, and otherwise animated clusters of cells we call animals. As those animals enlarged and started producing hard parts like shells and skeletons, the fossil record began playing tunes the average paleontologist could really appreciate.

Until recently, it has been nearly impossible to determine the ancient ups and downs of oxygen (and carbon dioxide) levels. Now, some interesting computer models that track the cycling of the elements carbon, sulfur, and iron as our dynamic earth huffs and puffs and thrashes around as continents drift and mountains rise and erode, have produced results that are consistent with fossil discoveries that show the shifting fortunes of life through geologic time. These models help elevate the role of plants from mere stage dressing to important playwrights in the drama of evolution. The illustration/graph I created for this article is a simplified version of one that appears in Peter Ward's book *Out of Thin Air* (Joseph Henry Press, 2006)

One plant revolution deserves special attention: The invasion of

dry land and the subsequent greening of land masses during the so-called Carboniferous Period of geologic time. Other plant revolutions followed, like the development of seed plants—especially the flowering kind, and the evolution of grasses, but the monstrous forests that flourished some 300 million years ago established the conditions that made vertebrate life on land possible. David Beerling provides a more detailed account in his book, *The Emerald Planet, How Plants Changed Earth's History* (Oxford University Press, 2007).

Sometime before 410 million years ago (in the Silurian period for you geology fans) green life latched onto the boulders and virgin sands of terra firma. As evidenced in the fossil beds of Wales, Scotland, and other locations, this land invasion also involved fungi—perhaps in symbiotic associations not terribly unlike today's lichens. Note that atmospheric oxygen levels rose to as high as 24% at this time—largely through the success of photosynthetic organisms in the oceans. These high oxygen levels may have made the invasion of land a viable option, not only for primitive plants and

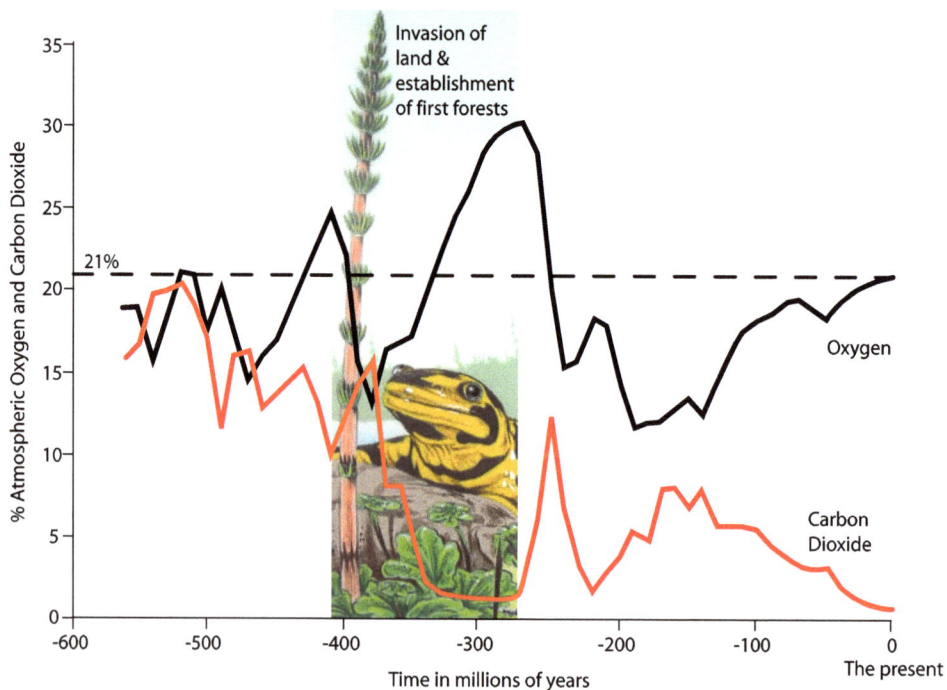

Invasion of land & establishment of first forests

% Atmospheric Oxygen and Carbon Dioxide

21%

Oxygen

Carbon Dioxide

Time in millions of years

The present

fungi but tiny arthropods as well—close relatives of springtails, mites, and other modern soil organisms.

But this first land invasion was only partially successful. No land vertebrates appear in the fossil record from this time period, for example. One reason for this according to Beerling may be that it took plants nearly 40 million years to develop a critical organ: the leaf. He believes the impetus for the development of stomates—the pores on leaves that open and close to regulate uptake of carbon dioxide—didn't occur until carbon dioxide levels began crashing over 360 million years ago. The need to maximize carbon dioxide uptake helped to spur the development of the flattened green solar panels that jumpstarted the evolution of vascular plants and led to the worlds' first forests.

Oxygen spiked at near 30% in the atmosphere some 300 million years ago as forests flourished. Paleontologists believe this not only allowed our vertebrate ancestors to make the transition from water to land (by augmenting oxygen uptake through the skin of amphibians, for example), but also encouraged "giantism" in insects and other creatures that occupied these primeval forests—like dragonflies with 2-foot wingspans and millipedes some 6 feet long. These oxygen levels crashed during the granddaddy extinction event of all time at the end of the Permian (another interesting story), but they eventually crept back up to their current baseline level of 21%.

The careful atmospheric records of carbon dioxide kept since 1950 (not visible on this graph), show, in addition to the overall sharp rise in CO_2 indicative of global warming, regular annual up and down cycles that change with the seasonal swings in photosynthesis, especially in the northern hemisphere. The cycles mirror the respiration of our green neighbors, the gentle breaths of a global biosphere.

So, have you thanked a green plant today?

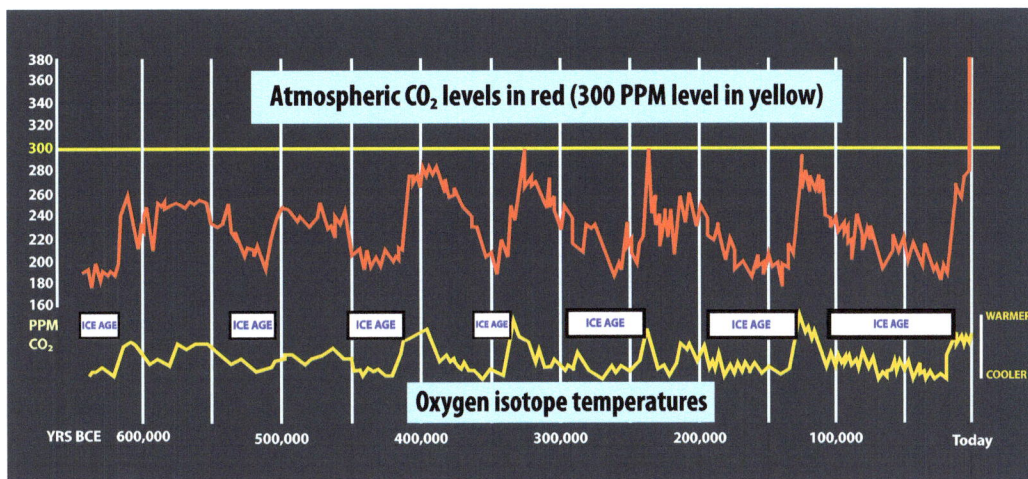

Atmospheric Carbon dioxide levels (red) and marine oxygen isotope levels (yellow). Both serve as proxies for global temperature (their curves closely match). Ice ages over the past 600,000 years are associated with troughs in CO_2 levels. Note the modern spike at far right, far above previous spikes over that period.

It's rather nice that most plants don't take too great an interest in the animal kingdom—except perhaps in science fiction stories. But who knows what another billion years of evolution might accomplish?

ARTICLE
Plants are aware, but do they care?

I read Daniel Chamovitz's book, What a Plant Knows, *and immediately recognized a kindred spirit. Chamovitz is a biologist quite aware that plants, animals, and every other living thing on the planet share a common biological heritage that is reflected in the perceptions and behaviors we are individually capable of. Since we share a roughly 50 percent common gene bank with bananas, we shouldn't be surprised that bananas and people draw on similar chemical tools and* modus operandi *when it comes to survival. Still, recognizing that plants can sense animals and to some extent manipulate what those animals do to their own advantage seems to fall under the category of "spooky."*

Since the behavior of predatory plants serves as a rather blatant example of animal-like behavior in plants (and because I thought it provided good visual candy), I decided to illustrate a Venus fly trap in the act of closing in on a fly. Like many of my illustrations, I used acrylic on matteboard but isolated the plant outlines with liquid mask so I could use a watercolor technique on the background.

As intelligent, social animals, we humans are great mind readers. We can detect love, lust, deception, concern, friendliness, and a host of other emotional states in the facial features and body language of our fellow primates with amazing accuracy. We get in a bit more trouble trying to mind read the intent of more distant relatives like dogs, raccoons, or armadillos. And as much as gardeners may love their plants, most draw the line at admitting to any meaningful emotional moment with such clearly alien creatures. Yet we share ancient genetic links with plants that show up in similar chemistries and methods of electrical communication. A few scientists even talk about "plant neurobiology," although the term is controversial.

Plants bend and dance to rhythms of light and gravity. They even smell danger and follow the scent of potential prey. Some species snap into action as part time carnivores. Others shrivel or sicken when touched. Many of the mechanisms used demonstrate our common evolutionary heritage as surely as if grandma had pasted their pictures in the family album. Daniel Chamovitz discusses many of these plant behaviors in his recent book, *What a Plant Knows*. Chamovitz isn't a big fan of the idea of a literal plant nervous system, but he does think that the term *plant neurobiology* is a good one for shaking up traditional thinking about what plants can and can't do. "What we must see," he says, "is that on a broad level we share *biology* not only with chimps and dogs but also with begonias and sequoias."

The hunters in us can appreciate a veggie that sniffs out other veggies and preys on them. Chamovitz carried out studies on the dodder, *Cuscuta pentagona*. This plant will die quickly unless it finds a host, like a tomato plant, to parasitize. Dodder seedlings, though noseless, can "smell" tomatoes (and other suitable hosts) and grow toward them in a lazy spiral until they make contact with a stem. The dodder then taps into the stems with needle-like microfilaments to obtain nourishment from their victim's sugar transport cells. Dodders can be fooled with cotton swabs dipped in a tomato plant mash. The most potent aromatic chemical in *eau de tomato* is beta-myrcene, although each plant emits a characteristic bouquet of additional compounds.

Plants usually emit odors for defensive purposes. Other researchers, for example, have worked out the details of how willow leaves injured by tent caterpillars release compounds that signal other leaves on the plant that they need to crank up production of phenolic and tannic compounds that discourage insect dining. Other trees—and perhaps other species—in the neighborhood may eavesdrop on these signals. "When a plant releases a smell in the air," Chamovitz asks, "is it a form of talking, or is it, so to say, just passing gas?" Whatever you call it, chemical detection in plants evolved because it has survival value.

Willows also demonstrate the role of common biochemistries among the spawn of planet Earth. When bacteria attack willows they cause the leaves to emit methyl salicylate, a close chemical relative of salicylic acid—and acetyl salicylic acid—the active ingredient in aspirin. Methyl salicylate bolsters the willow immune system and Bayer's product helps thin our

blood and get rid of headaches.

Plants don't have eyes, but they can sense light. Both plants and animals share primitive cryptochrome blue light photoreceptor systems that entrain our circadian rhythms—the same rhythms that give us jet lag when we cross several time zones. We share this basic photosensitivity with green single-celled brethren first evolved in Earth's early oceans.

Other light detection systems developed independently in animals and their green plant cousins. Plants never developed eyes with light sensitive rods and cones, but they do utilize phototropin receptors in shoot tips that allow them to bend toward a light source (phototaxis) and leaves possess red and far red detectors called phytochromes that control flowering times by responding to day length cues. If, as Chamovitz contends, "sight is the ability not only to detect electromagnetic waves but also the ability to respond to these waves," then plants can see.

Plants also sense the location of their various parts in relation to gravity's relentless tug. Chamovitz says that this proprioception is our real "6th sense" as animals. It's something we don't often notice unless we're drunk and our finger can no longer find our nose. Our sense of where one body part is in relation to another is not connected with one specific organ, although tiny particles in our inner ear contribute to balance and orientation in a gravity field. Plants also possess relatively heavy statoliths that provide growth cues to cells in both roots and the endodermis of aerial shoots. Plant statoliths evolved from modified chloroplasts (which were once independent microorganisms in the deep past).

Plants respond to touch. The leaves of so-called "sensitive plants" (*Mimosa pudica*) dramatically fold up at the stroke of a finger. Water rushes out of clusters of special cells strategically located so that when the cells deflate like empty water balloons the plant's leaves fold up. The chemical mechanism that depolarizes the cells' membranes and makes them pervious to water resembles the action potential of animal nerves.

Predatory Venus flytraps also demonstrate a sensitivity to touch that turns them into effective predators. The paired leaf lobes of a flytrap, you may recall, look a bit like open hamburger buns rimmed with nasty barbs. A few black trigger hairs stick up on the lobe's inner surface. If a bug is big enough to bang into two trigger hairs within twenty seconds, a nerve-like chemical depolarization occurs that allows water to rush out of certain cells so that the lobes snap shut and the barbs intermesh to become the bars of a fatal enclosure. When the insect dies, the plant obtains a nitrogen supplement.

But as much as plants demonstrate an eerie kind of parallel sensitivity to the world, we mammals still carry a unique cerebral filter on our shoulders that makes the experience different. "The independent paths of evolution," says Chamovitz, "have led to a uniquely human capacity, beyond intelligence, that plants don't have: the ability to care." He goes on to say "touch the leaves of an oak, knowing that the tree will remember it was touched. But it won't remember you. You, on the other hand, can remember this particular tree and carry the memory of it with you forever."

Words to ponder as we sit by winter fires waiting for our favorite green partners to turn their emerging shoot tips toward the coming spring sun.

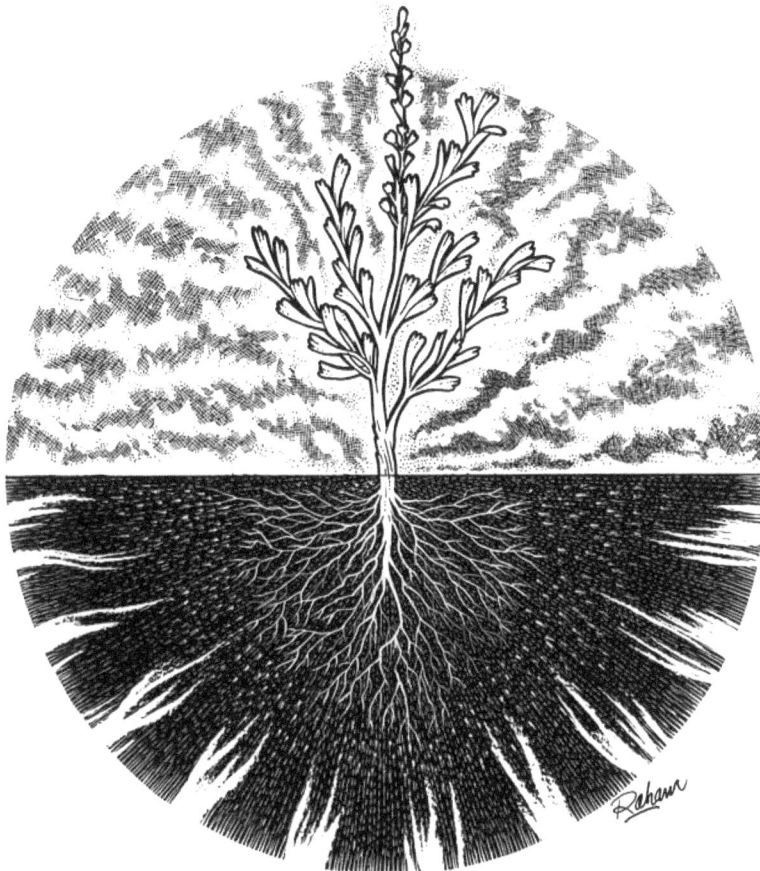

Plants like sagebrush (Artemisia) *wage war on other plants like tobacco* (Nicotiana) *by releasing potent chemicals like methyl jasmonate. The artwork was done on white scratchboard.*

ARTICLE
Pollen: not just something to sneeze at

I've been fascinated with pollen since my college days. I discovered pollen's potential for revealing the past during a course at the University of Michigan in 1969. We used a textbook called Pleistocene Geology and Biology *(which I still have), published in 1968. I don't recall the instructor's name, but she did a memorable job revealing how the intricate, beautiful pollen capsules called out the identities of their creators in time-specific detail. Periodically, new researchers "rediscover" pollen as a scientific tool, but I think they are first captivated by the endless variety of forms they produce—sculpted monuments built to last, but so small they remain invisible to all except those who choose to peer down the barrel of a microscope.*

The illustration to the left, "Milkweed Universe," was created for my first book, Dinosaurs in the Garden. *Milkweeds enlist bumblebees and other robust insects to transfer their pollen in saddlebag-like structures called pollinia. The scratchboard illustration on page 51 illustrates the kind of wind-borne pollen more likely to dampen your enthusiasm for plant sexual practices.*

This article appeared in the North Forty News *in March 2004.*

Sometimes it's hard to love something that makes your eyes tear up and your nose dribble, but pollen has so much going for it that it at least deserves our respect. Wind borne pollen, like that produced by grasses, tweaks my allergic responses the most, but neither you nor I would want to discourage the sex life of grasses. Grasses like corn, wheat, and rice feed the world. And pollen has turned out to be to be an amazing tool for a wide variety of scientific disciplines, including botany, paleontology, archeology, geology, and climatology. All this value is wrapped up in tough little packages the size of dust grains.

Male sex cells run amok

Plants practice an "overkill" method of reproduction because plant male cell producers (anthers) can't wander around looking for attractive ovaries in bars and grocery stores. Thus, a plant will produce hundreds of millions of pollen grains per flower and either coerce animals into helping them transfer that pollen or cast their fate to the winds by producing veritable clouds of particles, only a tiny fraction of which will ever surf an air current to a female flower of the right species. Here in the Rockies, Pine trees often coat nearby lakes with thick films of their wind-borne, amber pollen.

Insects, birds and various mammals, tempted by pollen's rich food value, can easily be conned into helping plants. Pollen grains are 16-30% protein, 1-10% fat, 1-7% starches, low in sugar, and high in vitamins. Plants that target animal messengers produce a thick, non-volatile oil that adheres globs of pollen to each messy feeder. The animals carry that pollen to the next flower they visit, usually one of the same species.

Wind borne pollen, light and dry, relies on blundering into the female pistil of the same species. Once at a female stigma, the male sex cell must send out a pollen tube which chemically digests its way through the neck of the ovary to the female egg cells. Two nuclei then traverse the tube. One nucleus fertilizes the egg cell, the other combines with a secondary female cell to form the endosperm that makes up most of the resulting seed--a food source to get the seedling up and growing.

A nearly indestructible time capsule

When pollen cell nuclei reach their destination, the pollen grain has served its biological purpose. Its vacated shell holds no more value to the plant than an empty soda can has for us. But to scientists that shell--because of its durability and uniqueness to its parent plant--is a time capsule packed with valuable information. And because pollen is so small and produced in huge quantities, it piles up in sediments everywhere.

For example: Drill into old lake or bog sediments, pull up a core of material, and you will find a continuous record of the rain of pollen (and spores) from all the plants that ever lived near that lake for thousands--even millions--of years. Scientists have done just that ever since Axel Blytt (1843-1898), Rutger

Sermander (1866-1944), and E. J. Lennart von Post (1884-1951) first came up with the idea. The ebb and flow of forests and prairies in north America and Europe have been read and recorded and used to measure the advance and retreat of glaciers as ice ages have come and gone over the past million years or so. An entire science--called palynology--has developed around all the nuances of interpreting pollen assemblages in the fossil record.

In 1954 one oil company took out a patent on a method based on the discovery that ancient ocean shorelines could be pinpointed by observing the size ratios of pollen in marine sediments. Smaller and lighter pollen grains tended to be transported farther away from shore than larger, heavier grains. Ancient shorelines provide geologists with clues to the location of oil deposits.

Pollen and ancient detective work

Pollen also infiltrates into some unusual places, leaving a unique signature that aids archaeologists. In late January of this year, *Nature* magazine reported that a French scientist, Serge Muller, is using pollen to determine exactly where certain ships lost at sea were built. It seems that pollen tends to stick to the resin used to seal a boat's hull. Therefore pollen produced by plants near a shipyard will provide a unique "birth certificate" for any boat built there.

Muller has been studying the *Baie-de-l'Amitie*, a ship that wrecked off the south coast of France some 2000 years ago. He believes builders sealed this boat's hull somewhere east of Italy because it contains both wood and pollen from *Platanus*, a tree that is restricted to the eastern Mediterranean. His conclusion is also supported by pollen from weeds like *Haplophyllum*, most species of which are also restricted to the same area. Such information helps unravel the mysteries of ancient commerce. Another researcher hopes to use Mull-

RAHAM

er's technique to figure out where the Persians built the fleet with which they invaded Greece in the fifth century B.C.

The human connection

Scientists have recently uncovered another interesting link between humans and pollen: an amino acid called amino butyric acid (GABA), which is an important chemical allowing human brain cells to communicate, also promotes the growth of pollen tubes necessary for plant reproduction. Perhaps the Navajos instinctively gave pollen its proper place in the scheme of things. Stephen C. Jett in "Navajo Wildlands" said "The symbol for life and productivity, for peace and prosperity, is pollen. Pollen symbolizes light."

Gesundheit.

ARTICLE
Lichens & the
hidden architecture of the living world

Lichens, those amazing liaisons between algae and fungi, do far more than dress up rocks with their bold colors. They should remind us that life has been around a very long time, that organisms do what they must to survive, and that survival depends as much on collaboration as competition. Lichens also demonstrate the fact that evolution proceeds utilizing the motto "If it ain't broke, don't fix it." Lichens persist as successful pioneers that will undoubtedly decorate the last human gravestones and crumbling artifacts.

This article allowed me to revisit lichen biology after an initial flirtation as a young science teacher. It also provided the opportunity to incorporate my passion for paleontology by using a painting of a Devonian scene that I had wanted to create since learning about Prototaxites, a tree-like fossil now believed to have been a "humongous fungus." In the painting, I clothed the fungal columns in lichens and placed them along a foggy shoreline.

Later, I submitted the painting to Focus on Nature XII, an exhibition at the State Museum of New York in Albany in 2012 and it received a Jury Award. It's always nice when you can captivate someone else with your vision.

The article originally appeared in the Summer 2010 issue of Colorado Gardener *magazine.*

Lichens grow where green plants can't. They latch onto bare rocks and slowly tear them apart. They crown mountaintops in red and gold and brazenly bathe in ultraviolet light. They cling to boulders in intertidal zones and survive salt spray and immersion. They speckle gravestones with splotches of orange, yellow, and gray. In Antarctic deserts they insinuate themselves beneath transparent lenses of quartz in defiance of temperatures that plunge far below zero Fahrenheit. They live, humble and unnoticed, in every terrestrial habitat, but they exemplify a basic, primitive, living, cooperative effort more pervasive than we can easily imagine.

Lichenologists, scientists who study these composite organisms, may smile and describe them as fungi that have discovered agriculture. A fungus seeks out an alga (single-celled green plant) or photosynthesizing bacterium (once called blue green algae) and builds a unique body called a thallus. We call boulders dressed with lichens "moss rock," but that's a misnomer. Mosses are more complex green plants, often growing near lichens on soils—and sometimes rocks. But scientists are beginning to realize that the lichen partnership (symbiosis) is an echo of more extensive associations in the green architecture of our living world.

More than forty years ago I became fascinated with lichens after moving to Colorado. Colorado's abundant rocks serve as host to many colorful species that paint stone with living patches of brilliant yellow, rust, and shades of green that vary from pale tea to slap-your-eyes chartreuse. Some lichens grow as crust-like patches. Others look like shrunken and dried lettuce leaves. Beard lichens hang from tree limbs like faded green strands of tinsel. But their appearance didn't captivate me as much as their biology. The cross-kingdom union of fungus and alga to produce a unique, totally different creature seemed like a science fiction writer's improbable ALIEN. And these lichen aliens show more survival potential than either humble partner alone. In addition, their 20,000 plus species produce 600 organic compounds found nowhere else in nature.

So, as writers do, I wrote an article for fellow science teachers and even scripted several filmstrips—an ancient teaching device some of you silverbacks may remember. I snapped photographs and drew some illustrations. I studied what science knew about these chimeras and marveled at what remained unknown. Then I moved on to other things. But recently I was pleased to run across the work of a new generation of scientists, including Francois Lutzoni (Duke), Suzanne Joneson (University of Idaho), and Robert Lucking (Field Museum of Chicago), among others. All three are studying lichen biology and evolution using newly available tools for decoding their genetics. In addition, I learned about an amazing fossil or-

ganism discovered in Welsh and German rocks that lived some 400 million years ago called *Prototaxites*, a fungal nightmare that stood 8 meters (26 feet) tall, dominating certain moist Devonian habitats at a time when bare land was the newest frontier of life.

Here's a third piece of information that I found quite interesting from a paper authored by Lutzoni and colleague A. Elizabeth Arnold of the University of Arizona: all the plant species that scientists have examined contain "lurker" fungi that grow inside their leaves, stems, or other above-ground tissue. Scientists have known for many years that fungi share important symbiotic relationships with many tree species, coiling around delicate rootlets, providing water and mineral resources otherwise unavailable in exchange for sugars, but the role of more pervasive fungal lurkers in all plant tissues remains unexplained. In my mind, lichenologists, reading the DNA barcodes of lichen fungi, and paleontologists, observing how fungi built tree-like monuments in foggy Devonian wetlands, are both building a case for the pervasive role fungi as a kind of superstructure for much of the green world we admire. Some scientists are looking into the possibility that lurker fungi also provide crucial services to green plants—perhaps by providing defenses against less domesticated, disease-causing fungi.

Fungi, without chlorophyll themselves, cannot produce sugars directly from the sun as green plants do. Typically, fungi serve as part of the cleanup crew on our planet, decomposing dead bodies and other detritus, but lichen fungi (and perhaps many other species) partner with green plants to survive. The partnership evolved over time—most likely a compromise when a fungus found itself unable to completely parasitize a green host. Some of the new research on lichens is beginning to clarify the complex way that lichen fungi and photosynthetic algae and bacteria find each other and establish their winning relationships.

When I first became enamored with lichens, it wasn't clear just how specific the union of a particular fungus with a certain alga might be. Now scientists know that 17,000 species of fungus compete to woo a few hundred species of algae and cyanobacteria (photosynthesizing bacteria). The lichen fungi turn out to be "needy aggressors" in the relationship. Their thread-like filaments called hyphae initiate contact with appropriate algae and branch profusely to entangle algal cells. But lichen fungi are almost never found living independently. Lichen algae can survive alone, but can't live in the harsh environments conquered by lichens.

And when did this all start? Lichen fossils have been found with the early land coloniz-ers entombed in the Rynie Chert of Scotland, nearly 400 million years ago, but researchers like Lutzoni and Lucking believe lichen partnerships may go back some 600 million years. Lutzoni's work with lichen fungus DNA is beginning to nail down relationships and suggest at timelines.

Surprising fossils like *Prototaxites* show that fungi, primitive photosynthesizers, simple green plants, and tiny arthropods formed complex networks of relationships from the very beginning that allowed for the initial colonization of dry land. Today the "lurker fungi" showing up in the flowering plants we all know and love hint at an ancient architecture of relationships that still holds the modern world together with the delicate transparent threads of fungal hyphae.

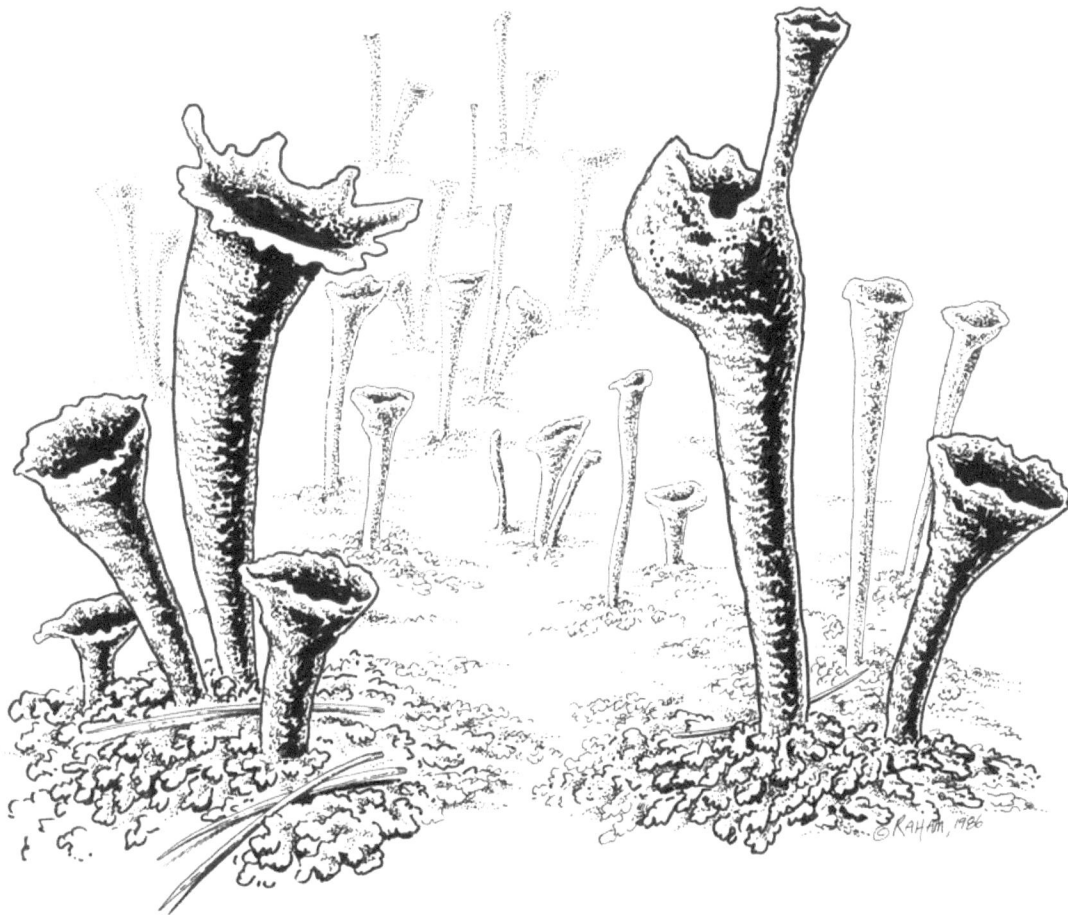

Foliose lichens look like a bizarre, Lilliputian forest seen up close and personal.

A photo of Margulis in full lecture mode and an illustration created by her son, Dorian Sagan, for her book, Five Kingdoms: An Illustrated Guide to the Phyla of Life on Earth, *inspired this scratchboard illustration. The book, co-authored with Karlene V. Schwartz, enjoyed several updates and revisions. It demonstrates the power of illustration in clarifying otherwise abstract or even obtuse scientific jargon.*

The five fingers of the hand represent the five major kingdoms into which microbial life evolved over Earth's 4.5 billion year history. Life, in conjunction with the physical elements that comprise Earth, manifests itself as a self-regulating planetary ecosystem referred to by James Lovelock as "Gaia." This "mother goddess" label inspires a kind of "worship" by some environmentalists that more conservative biologists feel may prejudice its scientific value as a theory.

ARTICLE
Lynn Margulis: Champion of the microcosmos

The world lost a genius when Lynn Margulis passed away in 2011. During her lifetime she angered many people, as those who slap important faces and force them to see the world with a new clarity are apt to do. And like other world changers, she pointed out human hubris and self-absorption while dissecting staid assumptions with the sharp scalpel of a childlike sense of wonder and observation.

Margulis' revolutions started in the academic halls of life science buildings and took decades to overthrow traditional views. For some, Margulis hovered on the fringe of academic acceptability at a time when her very gender brought her radical notions under suspicion. Besides, she began married life with Carl Sagan, a brilliant astronomer also academically suspect because he catered to the "masses."

At any rate, I owe much of my biological perspective to Margulis' work, as you will discover in this article and others. This piece appeared in the education issue of Colorado Gardener *in 2012.*

Nicolaus Copernicus (1473-1543) took away our position at the center of creation. Charles Darwin (1809-1882) showed that we are just animals with inflated brains. Lynn Margulis (1938-2011) revealed during her lifetime that we are merely a colony of microbes with an attitude. Dr. Margulis' legacy is yet to be fully realized, but her work has already transformed the way scientists think about the living world, from basic biology to the intricate mechanics of a properly functioning biosphere.

I was particularly struck by Margulis' contributions while writing a science book for middle school students called *Biodiversity and Food Chains*, due out from Chelsea House this coming spring. Many of the topics I covered were either absent when I went to school or filtered through a more clouded lens of understanding. Margulis, just eight years my senior, began transforming what science thought they knew about biology, when I was in college during the 60s. E. O. Wilson, who invented the term biodiversity to describe Earth's varied life forms, and no creative slouch himself, referred to Margulis as "one of the most successful synthetic thinkers in modern biology."

Margulis realized, and then went on to help

prove, that all complex creatures—essentially everything big enough to see with the naked eye—are really composite mega colonies of microbes. Microbes, in fact, have defined life on Earth for most of the last 3.5 to 4 billion years. "Do historians begin the study of civilization with the founding of Los Angeles?" Margulis asks. "This is what studying natural history is like if we ignore the microcosm." Not only do microbe cells outnumber human cells 10 to 1 in our bodies, each human cell is itself a composite construction hammered together eons ago during a microbial collaboration Margulis calls symbiogenesis.

Margulis observed that the cells of complex animals are actually chimeras, or assemblages of several organisms. The proof lies in the DNA. In human cells, for example, the DNA that defines our architecture and development resides in the cell nucleus. However, the energy-producing cell organelles called mitochondria that crank up our metabolism have their own unique DNA that once belonged to an independent organism more than two billion years ago.

Margulis proposed this idea in 1964 in a paper entitled "On the Origin of Mitosing Cells." Over a dozen journals turned down publication. Finally, some courageous or perceptive editor at the *Journal of Theoretical Biology* published the work in 1967. Its debut garnered requests from over 800 fellow scientists, stunning the biology department at Boston University where she taught at the time. Subsequent research over succeeding decades transformed the concept from wild idea to an accepted tenet in science.

Several of Margulis' more than fifty books have been collaborations with her son, Dorion Sagan. Dorion's famous astronomer father, Carl Sagan, explained macro science to great effect with his TV series Cosmos. Dorion and his mother attempted to reach popular audiences with books like *Gardens of Microbial Delights* (1988) and *Micro-cosmos* (1986).

Dorion also contributed a sketch (with design contributions from Donna Repard) that was used on the cover of the third edition of *Five Kingdoms, An Illustrated Guide to the Phyla of Life on Earth*, that Margulis co-wrote with Karlene V. Schwartz. This graphic, which I reproduced in my portrait of Margulis, is meant to show that Earth is more than "the third rock from the sun," but an integrated living system with organic and inorganic parts. Margulis, in fact, worked with scientist, James Lovelock, in his also controversial "Gaia theory" that the Earth as a whole functions as a kind of "biostat" keeping our planet in a balance that promotes the survival of its living components.

While Darwinists and Neo-Darwinists emphasized the competitive aspect of natural selection in the process of evolution, Margulis and Sagan emphasize symbiotic relationships between living things. "Life did not take over the globe by combat," they wrote in 1996, "but by networking." A colleague and former student, Margaret McFall-Ngai, also wrote "The fact of the matter is that everybody is waking up to (the importance of symbiogenesis), and it's only been going on for about the last ten years, so basically Lynn was way, way ahead of her time."

On the practical side, her insights have permeated all aspects of the biological sciences. Doctors must treat the human body as the colony of walking bacteria it is. Ecologists are realizing that if they poke Gaia in one place, she may spasm somewhere else, and horticulturists have to realize that weeds are not just misplaced plants, but entire transplanted communities of microbes eager to put down firm roots—and perhaps delicate fungal threads.

Thanks for the new insights, Lynn, even if they may come as yet another blow to our collective human egos.

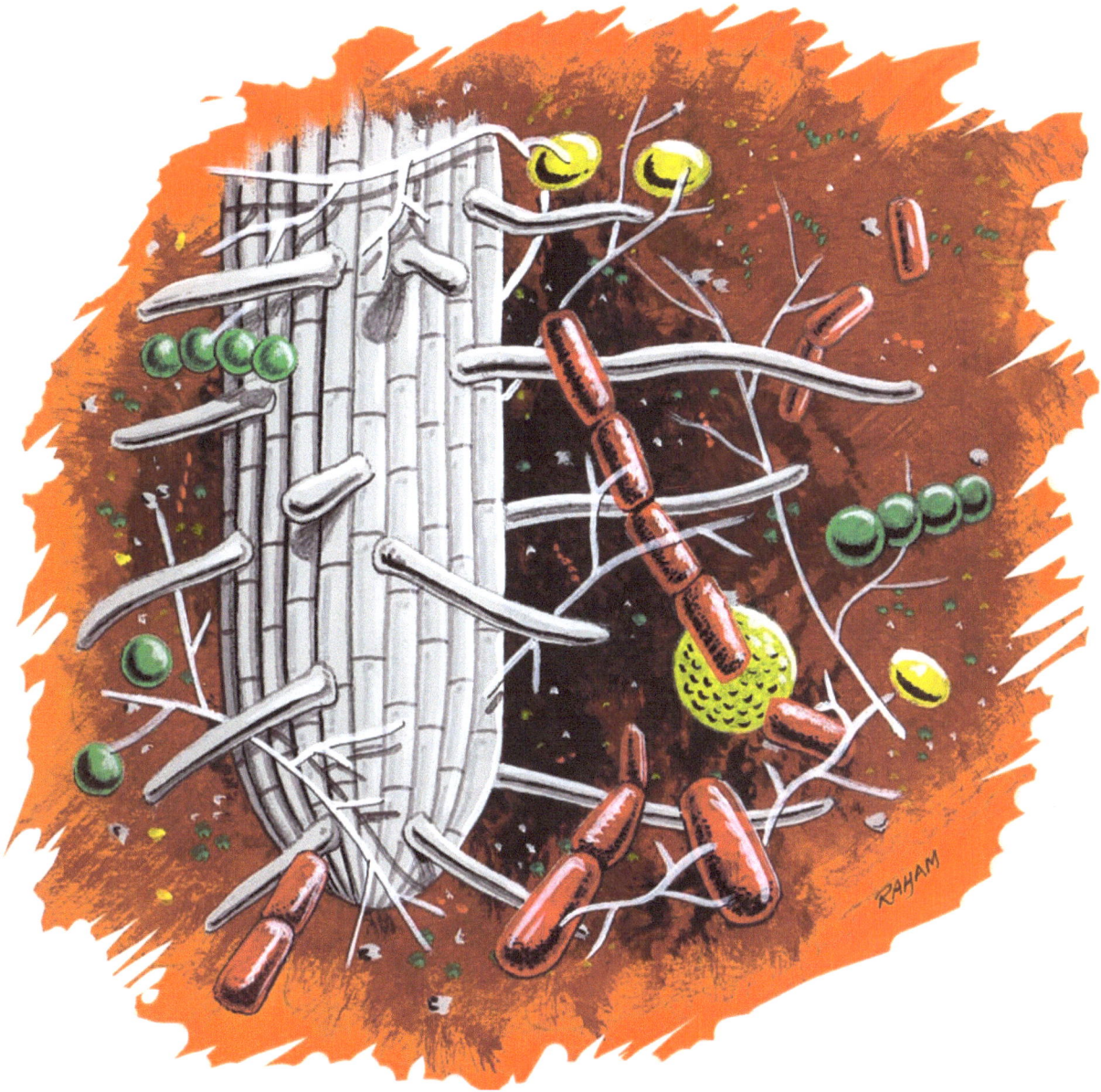

I created this illustration to accompany an article called "Turmoil over soil, a climate-warming wildcard" in the North Forty News. Soil holds three times more carbon than the atmosphere does. A host of fungi, bacteria, and tiny invertebrates preside over the decay of organic matter and form various liaisons with the green plants whose root hairs invade their domain. Margulis made us more aware that microbes are the latticework that holds our familiar macroscopic world together.

ARTICLE
The garden microverse

I confess that this article represents a montage of journalistic odds and ends—fascinating scientific investigations tied together by the diverse and talented members of the micro-cosmos involved. As I mentioned in the introduction to the previous article— my ode to Lynn Margulis—her insights continue to influence the way I look upon nature and its long, evolutionary history on our planet.

Evolution truly is the unifying concept in biology. Scientists continue to refine details, but the evidence of that creative process lies open for all to read in the tangled, redundant, and often imperfect guidebook of individual genomes. Microbes wrote the fundamental lines of that guidebook long before anything bigger than the dot at the end of this sentence emerged to add macroscopic addenda.

I liked working out the perspective for the scratchboard drawing and hope it conveys the notion that complex life is an edifice built by and supported by tiny living beings that, with Margulis' insights, should shimmer in our mind's eye like "points of light in a pointillist landscape."

This article appeared in the Spring 2009 issue of Colorado Gardener *magazine.*

The biologist Lynn Margulis once said, "Life on Earth is such a good story you cannot afford to miss the beginning." As a Microbiologist, Margulis knows that one doesn't find most life without a little magnification. Not only that, microbes set up shop so early on the planet, that all other forms of life are post-it notes stuck on a long-woven fabric of microbial relationships. As we dig, spray, transplant, and chemically stimulate our gardens, it behooves us to remember that we are shaking up a beehive of relationships that could sting us if we make poor choices.

It's all microbe chemistry

Every chemical reaction that churns away in you, me, Fido, and the Burpee crops behind the shed was first field tested by microbes BILLIONS of years ago. Sometimes they made serious blunders—for them, at least. When cyanobacteria first cranked up photosynthesis a couple of billion years ago, it seemed like a great way to harvest energy from the sun. But oxygen turned out to be a major poison that eventually frizzled most of the other microbes around at the time or forced them into secluded habitats.

No matter. Crises bring innovations. Some microbes burrowed into fledgling animal cells to become energy producing mitochondria—cell organelles no animal can now live without. Other microbes took up residence as chloroplasts in other creatures destined for greatness as trees, flowers, and other green plants. Bacteria now line our guts, keeping out interloper microbes (most of the time) and performing essential services like vitamin K production, in the case of *E. coli* bacteria in our large intestines.

Plants teamed up with the sometimes microscopic, sometimes macroscopic denizens of Kingdom Fungi. Much of this collaboration may have developed when plants broke the water barrier and crept onto dry land some 400 million years ago. But plants, too, have their longtime bacterial buddies and antagonists.

Learning about all these interactions may seem as pointless as an idle romp through Facebook, but everyone interested in a healthy and smoothly functioning biosphere needs to pay attention to these microscopic liaisons.

Those nodules look good on you

Most people are familiar with the nitrogen-fixing bacteria that form root nodules on leguminous plants like soybeans, peas, clover, and so forth. Bacteria form nodules on the roots of these mostly temperate plant species and busily convert atmospheric nitrogen to forms that the plants can use. When the plants die their remains make excellent fertilizer. Scientists have known for some time that the *Rhizobium* bacteria specializing in this plant fraternization carry "nod" genes that allow them entry to the plants through their root hairs.

In 2007, Science magazine reported the discovery of new strains of tropical bacteria (*Bradyrhizobium*) that perform similar services for tropical plants, but go one step better. The *Bradyrhizobium* bacteria produce a plant hormone mimic similar to cytokinin that allows them to enter through cracks or wound-like areas in the plant. The bacteria set up shop in nodules on the stems of the plants rather than the roots, which makes access to nitrogen more efficient. Scientists believe they might be able to coax these microbes to infect other leguminous tropical species, thus improving efforts to enrich soil that has been leached of nutrients through over farming.

Don't forget to invite trees to the party

South Americans learning how to rebuild the quality of farmland soil have also been discovering the value of reintroducing trees into the mix. One tree, *Erythrina* (in the same family as peas), sports a straight, slim trunk suitable as a bean vine pole but also carries a suite of nitrogen-fixing microbes. Neem, a member of the mahogany family, secretes leaf chemicals that sterilize insect eggs. Fruit trees interspersed among the beans, yucca, and corn add much needed organic material to the soil mix.

And forest tree roots often depend on the machinations of soil fungi to augment the ability of their roots to absorb water and nutrients. One of the more famous of such associations is that between various truffle fungi and the roots of oaks, hazelnuts, and other tree species. U.S. businesses like New World Truffieres, Inc. constantly seek to inoculate new tree hosts with the fungus to grow this tasty crop of mushrooms. Many other fungus tree associations await discovery.

Insect partnerships

Insects, too, owe their success and often survival to associations with microbes. According to the university of York Department of biology in the U.K., at least 10% of insect species bear intracellular microbes restricted to special structures called mycetocytes. This association is a true symbiosis. The microbes culture poorly, if at all, outside their insect partners.

Aphids provide a good example. They specialize in sucking the sap from plants (like your prize roses). Saps, though rich in carbs, yield few, if any, amino acids. When scientists treat aphids with antibiotics that kill the bacteria in their mycetocytes, the aphids fail to grow and reproduce. Other sap-suckers falling into this group include whiteflies, planthoppers, leafhoppers, and cicadas. Insects that live on vertebrate blood like tsetse flies and sucking lice also need microbial helpers. Add various beetles and all cockroaches to this list.

Soil: the living substrate

The examples given above only hint at the microverse of relationships that thrive in every handful of soil. Primitive arthropods called springtails, each the size of a grain of pepper, jump and burrow in their countless millions, scavenging organic detritus as they've done since plants first conquered dry land. They cavort with mites, pseudoscorpions, and other minute arthropods all jumbled in a threadwork of fungal hyphae and mineral grains. Translucent roundworms, both in and out of the soil, are so numerous as parasites on all sorts of living things that their remains alone would form a ghost image of the macro universe we're used to observing if all other living things disappeared. And while a "living soil" might at first blush seem scary to some, a lifeless soil would be scarier still, because it would portend the end of all us macroorganisms with delusions of grandeur.

Each of the soil's tiny creatures, most likely, has their own collection of microbe partners. E. O. Wilson, the entomologist/ecologist and inventor of the term biodiversity, says we have catalogued perhaps 1.8 million species on earth, but as many as 10 million await discovery. Most of them are microbes.

This view of a deeply integrated biosphere serves as both a source of wonder and beauty. As Margulis said, "The world shimmers, a pointillist landscape made of tiny living beings. Giant redwoods and whales, mosquitoes and mushrooms are intricate symbiotic networks, modular manifestations of the nucleated cell."

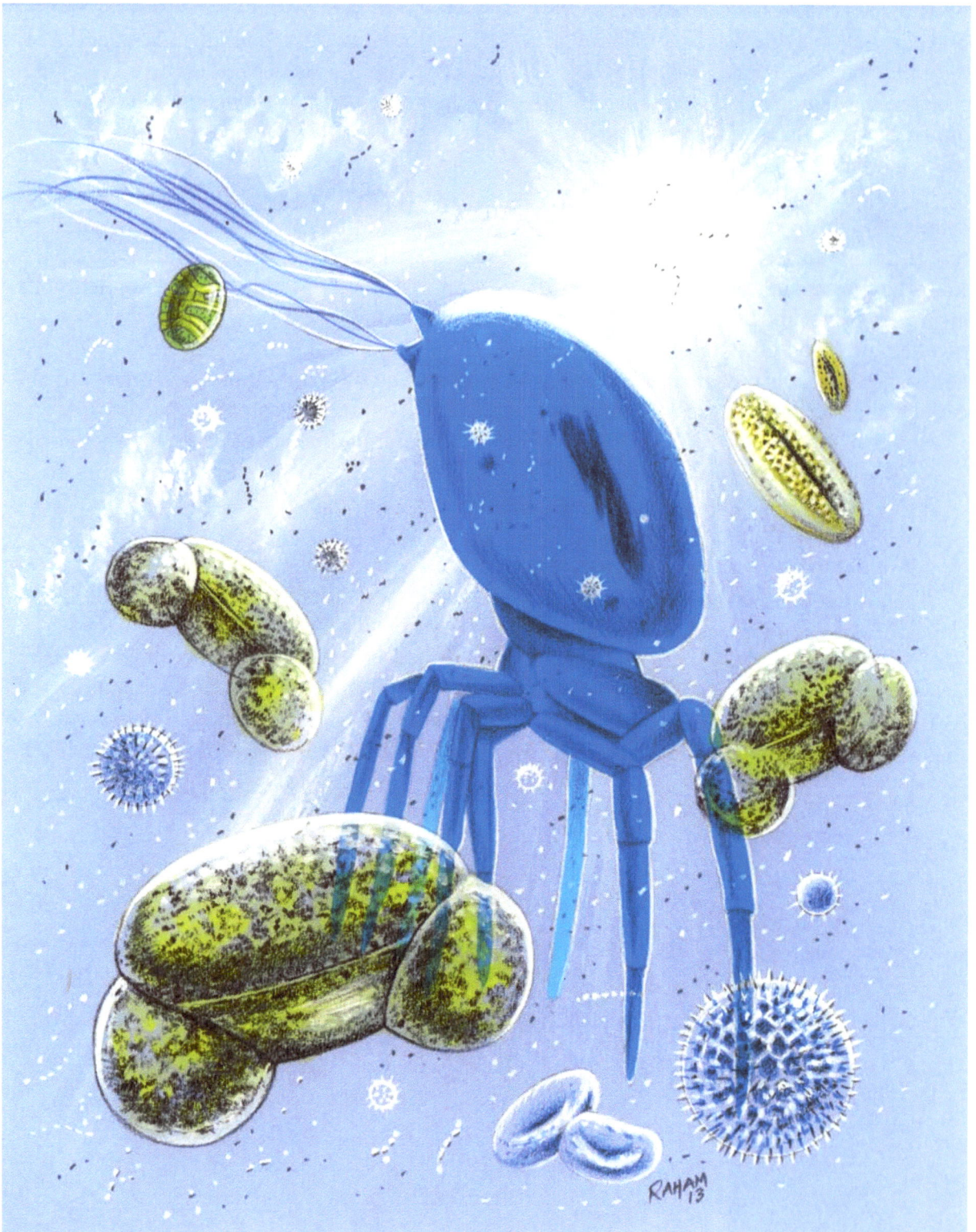

The Earth's atmosphere supports a wealth of life from transients like spiderlings adrift on gossamer threads of silk, fungal spores, and plant pollen to resident bacteria. In Colorado we might see Ponderosa pine pollen (three-lobed structure in fore and mid ground), milkweed pollen ("life savers" near bottom of the picture), and football-shaped Arabidopsis pollen (just to the right of the spider).

ARTICLE
Alive & Aloft in the Aeolian Zone

This article developed from several sources, including interviews with Paul DeMott at Colorado State University, Tom Hill at the University of Wyoming (both atmospheric scientists), William Bryant Logan's book about life aloft, and what I knew about the rebirth of the volcanic island of Krakatau in Indonesia after its eruption in 1883. In many ways the work done by these scientists and others reflects the changing view of the microscopic world pioneered by Lynn Margulis in the 1960s. (See my earlier essay, "Lynn Margulis: champion of the microcosmos.")

I submitted this piece for award consideration with the Colorado Authors' League and was pleased when it received recognition as Best Feature Article/Essay in 2014. It originally appeared in the April 2013 issue of Colorado Gardener *magazine.*

The illustration is mixed media on matteboard. I am constantly startled by all the beauty and diversity of the unseen worlds revealed by the microscope.

Breathe in. Exhale. Repeat as necessary. In the course of an hour you will have filtered a third of a cubic meter of air, extracting its vital oxygen to use in the slow burn of metabolism and venting carbon dioxide waste. You will also have inhaled roughly 1500 bacterial cells, hundreds of fungal spores, and a lusty sampling of pollen, tiny arthropods and other minute critters. This airborne life is, for the most part, a good thing. Really. The aerial ecosystem helps the planet distribute soil and moisture, heal the wounds caused by volcanic eruptions and other disasters, and keeps the body terrestrial functioning properly. The atmosphere truly acts much like our own circulatory system, providing a fluid highway for the chemical commerce that keeps us alive and well.

Our collective ignorance of this fascinating domain has been profound. Greek mythology created Aeolus, keeper of the winds. Greek sailors probably noticed that dirt, insects, plant parts, and a wealth of other things often rained down on them hundreds of miles from the nearest shore. In 1832, Charles Darwin aboard the *HMS Beagle* captured some of this airborne material 300 miles off the coast of Africa. Darwin sent his samples to the German microbiologist C. H. Ehrenberg who later sketched the wealth of bacteria, pollen grains, plant silica and other things in the naturalist's collection.

In the 1930s, scientists collected such "aerial plankton" in nets tethered to kites and later the wings of biplanes. In 1974, Russian rockets collected air samples at heights of 35 to 50 miles containing six species of common microbes. In the Himalayas researchers have counted a rain of 400 specimens dropping onto a 10-square-meter plot in 20 minutes—creatures air lifted from the plains far below. More recently, a team of scientists based at the Georgia Institute of Technology in Atlanta rode on NASA airplanes studying hurricanes. They managed to bag samples containing a cornucopia of bacteria, fungal spores and other microbes in the troposphere 10 kilometers over the Caribbean.

Like you and me flying coach to visit Grandma, many of these creatures are transients in the planetary circulatory system, but scientists suspect that certain bacterial species found in nearly all samples may reside there permanently, using airborne oxalic acid as a food source and reproducing on the fly. So, what is this circulating ecosystem doing in the grand scheme of worldly affairs?

1.) Life in the Aeolian Zone (Remember Aeolus!) may help determine when and where rain falls.

I recently interviewed Paul DeMott, an atmospheric scientist at Colorado State University, and Tom Hill, a biologist at the University of Wyoming. Both are sampling microbial life over Northern Colorado and the Midwest and learning how bacteria may help determine when and where rain falls. Rain begins as a kernel of ice forming around a speck of dust in clouds. Researchers are beginning to realize that much of that "dust" is really living bacterial cells that allow ice to crystalize at warmer temperatures and rain to fall more readily. (See "The Sky is alive with biological rainmakers" at http://www.northfortynews.com/the-sky-is-alive-with-biological-rainmakers/) Some biologists speculate that microbes may play a significant role in regulating global weather patterns.

2.) Micro aeronauts heal Earth's wounds.

Scientists have measured the rebirth of ecosystems after violent volcanic eruptions. In 1883, Krakatau in Indonesia erupted with the force of 100 to 150 megatons of TNT and reduced its parent island to the cinder relic called Rakata. In 1884, Belgian biologist, Edmond Cotteau, found the first returning life form: a spider who had landed using its parachute of silk. Some organisms rafted to shore or swam, but birds, bats, and insects came by air with hitchhikers: bacteria, protozoa, fungal spores, nematodes, mites, and seeds. By 1886, only three years after the eruption, 15 species of grass and shrubs had established homes there. In 1919, grassland surrounded patches of forest. A 1984-85 survey recorded 30 species of land birds, 9 species of bats, several rat species, 9 species of reptiles, 600 invertebrate species and a host of microbes.

3.) Earth's wind systems also transport soils from continent to continent.

The amount of soil hefted into the troposphere by cyclonic winds and transported by Trade winds is not trivial. Approximately 1.5 billion tons of soil travels from African and Asian deserts to the Americas each year. As William Bryant Logan observes in his recent book, *Air: The Restless Shaper of the World*, "Half the desert dust that blows around the world comes from Africa. " With the dust come precious loads of iron, calcium, and phosphorus. "Without this airborne dust," Logan continues, "there would be no rain forests in Brazil." Some of this mineral treasure also reaches the slopes of the Rockies and our own high country gardens.

Breathe in. Exhale.

Logan says, "Breath turns place into habitat. If you can breathe there, you can live there. The living have succeeded in occupying the entire air, from bottom to top—an area four times as great as all the water in the ocean—and over the full range of possible climates." All that life performs the slow, flameless burn of cellular metabolism, gobbling sugars, and igniting them with oxygen to make living tissue. Life keeps on keeping on.

As Logan concludes in his book, "We, and everything around us, live in the midst of a slow cold flame." The wind carries living embers to every corner of the planet, fanning the flame, and encouraging the next collective planetary breath.

I created this slime mold illustration just because I think slime molds are cool. They produce spores that contribute to life's aerial mixture. The spores germinate into amoeba-like cells that feed until supplies run out. Then, the single cells emit "come hither" chemical signals that cause the individual cells to congregate into the spore-bearing structure depicted here.

R. GARY RAHAM

ARTICLE
The Mysteries of Metamorphosis

Biology provides endless opportunities for amazement—especially trying to figure out how life took a limited toolkit of genetic information and profoundly exploited it over millions of years in the interests of survival. In other articles I've talked about various examples of symbiosis, communal living, and evolution, but this piece about metamorphosis brings into question a basic tenet of life science: that species are defined by their ability to breed true. If the ideas of Donald Williamson prove to be correct, bizarre liaisons between genera, orders, or even phyla may sometimes result in the hijacking of entire suites of genes. Moreover, some of the evidence for this revolution may come from a process we all marveled at as children: the astounding transformation of a caterpillar into a butterfly.

The illustration depicts the metamorphic stages of a Mourning Cloak, Nymphalis antiopa. *They inhabit streams, open woodlands, and urban parks. Trees like willows, birches and cottonwoods serve as host plants. I sketched the piece with colored pencils on gray paper and struggled at first trying too hard to make it photo realistic. When I loosened up my technique a little—perhaps succumbing to some mental transformations of my own—the rest of the illustration progressed quickly to something I (and my editor) liked.*

This article appeared in the Harvest issue of Colorado Gardener *magazine in 2014.*

The Mourning Cloak flutters aloft on wings flashing glimpses of terra cotta, purple, and white. The butterfly detects an alluring scattering of molecules in the air that he follows unerringly. He meets and mates with a paramour who then leaves eggs to hatch on blades of aromatic rabbitbrush. Each egg yields a ferocious, worm-like larva that feeds and molts and grows. Eventually, each surviving youngster pauses and transforms. Within a horned and knobby chrysalis, Chromosomes puff in distinct patterns; DNA unwinds. Old tissue dissolves and reforms. A new adult begins to twist and stretch and will soon emerge from its ruined pupal skin sampling the air for that certain irresistible scent…

Sounds like the stuff of science fiction, but butterflies and other insects routinely practice such miraculous metamorphoses—often readily available in our gardens. As children, we usually watch the drama unfold at least once after we encase a found cocoon in a bottle. Some scientists never lose the sense of wonder that transformation engenders. The Frenchman, Henry Fabre, most often comes to mind for his elegant behavioral observations. But Darwin studied insects carefully and pondered the metamorphosis miracle. Vincent B. Wigglesworth spent a career unraveling some of the crucial hormone chemistry. Even less familiar names include American entomologist Carroll Williams and husband and wife Lynn Ritchford and James Truman.

In his recent book, *Metamorphosis*, scientist and author Frank Ryan described the decades of scientific sleuthing that created our current understanding of how these transformations occur within species and the evolutionary advantages that accrue for insects and many other creatures that produce young that don't directly compete for food and space with adults.

Ryan also introduces a modern scientific rebel who wants to change the story: Donald Williamson. His version seems even more science fictional. He claims that the genes involved in metamorphosis—in at least some cases—may represent cases of ancient hanky-panky across not only species boundaries, but sometimes across the biological gulf between genera or even phyla of organisms. Only time—and the dedicated obsessions of another generation of scientists—will determine if his ideas represent fact or fiction.

Williamson spent his working life as a planktologist at the former Marine Laboratory on the Isle of Man. Planktologists spend their days looking at the myriad microscopic creatures that live on the surface of the ocean. Tiny plants produce most of the atmosphere's oxygen. Tiny animals form the base of a vast food chain that ends with whales and sharks and King Sooper's shoppers. And many of those tiny animals represent the unlikely-looking larval young of starfish, crabs, and other invertebrates that we could readily identify in their adult forms. In some cases, over the course of his long career, Williamson observed larval similarities that he finally decided couldn't be explained by the normal mechanisms of evolution.

Although there are too many details to relate in the 800 words allotted here, the essence of Williamson's argument is that although hybrid unions are exceedingly rare, sometimes they happen—and on even rarer occasions they result in viable offspring with long strings of hijacked genes they may explain larval similarities across what were once considered impregnable barriers of inheritance.

Williamson performed some experiments with marine invertebrates that seem to support his claim, but revolutionary theories require exceedingly convincing proof. Williamson also expanded his theory to include the radical transformations of butterflies and other insects that display complete metamorphosis. He thought perhaps some union occurred between ancient worms and primitive insects some 400 million years ago. Although he had no experimental evidence for this, Williamson contended that new techniques in quickly reading the DNA blueprints of various organisms might provide the proof.

The beauty—and success—of scientific discovery depends on making hypotheses capable of being disproven by experiment. Colorado entomologist, Paul Opler, when told about Williamson's ideas said, "This concept just goes against everything I've ever been taught. This doesn't mean that it's not true, it's just an unprovable hypothesis, and it can't be scientifically falsified." For now at least, Opler's view represents mainstream scientific opinion.

But other cases of gene hijacking once thought radical have become part of mainstream biology. We now know that the cellular machinery called mitochondria were once free-living bacteria that formed an unlikely union with other cells. Chloroplasts in plants were once independent cells with the ability to turn sunlight into chemical energy. So, just possibly, the seemingly miraculous process of metamorphosis may be another example of an ancient "dangerous liaison" that proved valuable to survival.

At any rate, should you find a butterfly in the garden—at any stage in this process—take a moment to revisit the mystery and wonder that captivated you as a child. Metamorphosis could just be a mystery more astounding than anyone imagined.

In adult form, hawkmoths hover like hummingbirds over long-necked flowers and tap nectar supplies with their long proboscises. As larvae, the gardener knows them as hornworms, like the common green tomato hornworm.

ARTICLE
Great Blue "Dinosaurs"

I wrote "Great Blue Dinosaurs" in 1990 at a time when scientists were just nailing down the bird-dinosaur connection with new fossil discoveries and a new appreciation for some observations made in the early days of paleontological discovery. Subsequent discoveries in Northeastern China of Microraptor, Caudipteryx, *and other feathered dinosaurs (with the feathers dutifully preserved) firmly fused the dinosaur-bird link. In some cases, where pigment-containing melanosomes are preserved, it's even possible to guess quite accurately about feather colors.*

In Colorado, it's fun to watch great blue herons poised to spear a fish in quiet waters or see their characteristic form soaring overhead. The writer Loren Eiseley in his book The Night Country *tells of falling asleep on a western field trip only to awake to the "cold, yellow-eyed stare" of a blue heron curious about the sleeping mammal below him. "He was standing quietly on one foot and looking," said Eiseley, "like an expert rifleman, down the end of a bill as deadly as an assassin's dagger."*

It's appealing to know that some dinosaurs survived the big extinction. We can only hope they hold no lasting grudge against the furry mammals who took their place and often cage some of their descendants as feathered songsters and eat others as Chicken McNuggets®.

"Great Blue Dinosaurs" appeared in From the Canyons, *a publication of the Canyonlands Natural History Association.*

The illustration, pencil and acrylic on colored matteboard, shows a heron's dinosaur reflection in the water.

Follow those tracks near the water. If you are careful and quiet enough, they will lead you to the "dinosaurs" that made them. Don't expect fossilized beasts swept into some oxbow turn of the river a hundred million years ago. Instead you will find living animals that wait patiently or stalk carefully until they can spear a passing fish with a deadly thrust of their beak. If you come upon them suddenly, they may take to the air with a "kraak" call that echoes off canyon walls along with the sound of the slow, powerful strokes of their wings.

These particular dinosaurs are usually called great blue herons. Many scientists are now quite confident that herons and all modern birds are direct descendants of some small, theropod dinosaurs. In silhouette, at sunset in some remote canyon, the great blue heron could be newly emergent from some distant past as it walks slowly through the water or stands at rigid attention, waiting for a careless fish to mistake a predator's shadow for shelter.

Up close, this bird is a striking creature with long legs, compact body, and elegant plumage. The great blue heron sports a white crown, cheeks, and neck. Black stripes on either side of the crown merge at the back of the neck to form a long, dark crest. The thighs are a distinctive chestnut color. During mating season, long plumes adorn the back and lower neck. The predominant gray-blue color of most of the bird gives it its common name. None of these visual characteristics, however, serve to link herons with their dinosaur ancestors. The relationship between dinosaurs and birds is forged by information based on what the dinosaurs left behind—mostly bones.

These bones tell us that there were "lizard-hipped" and "bird-hipped" dinosaurs. The lizard-hipped variety contained both ferocious predators like the tyrannosaurs and vegetarians like *Apatosaurus*. The bird-hipped sort, vegetarians like *Triceratops*, grazed with beak-like mouths and teeth adapted for chewing. The hips of this latter group were superficially more like those of modern birds, but that was the only similarity. Birds, it seems, developed from small "lizard-hipped" predators whose skulls and legs provide the best evidence of their common lineage.

We aren't used to thinking of small dinosaurs, but there were many of them. The Jurassic-Age Coloradoan, *Ornitholestes*, stood six feet tall and weighed about fifty pounds. *Ornitholestes* was one of many small predators

called coelurosaurs that may have hunted in packs like wolves. *Compsognathus*, a chicken-size dinosaur known from well-preserved fossils in Germany, currently holds the record for smallest dinosaur. Many of these diminutive dinosaurs shared bird-like characteristics. They walked on their hind legs, had hollow bones, large eyes, supple necks and bird-like feet.

Skeletons of *Compsognathus*, in fact, have been confused with the remains of *Archaeopteryx*, the missing link of bird genealogy. *Archaeopteryx* stood as tall as this dinosaur cousin, lived in the same arid island habitat, but possessed wings covered with flight-modified feathers and the fused collarbones we usually call "wishbones." But *Archaeopteryx* also sported an unbird-like boney tail, sharp teeth housed in a reptile-like skull, and wrist bones reduced in size but not fused as in modern birds.

Had you been wandering the Jurassic island home of this "almost bird," however, the tracks on the sand would have looked quite familiar—perhaps not unlike great blue heron tracks. Both small coelurosaurian dinosaurs like *Compsognathus* and early birds like *Archaeopteryx* have the three toes forward and one toe back pattern of modern birds. This distinctive feature along with other details of the foot, ankle, shins, and hips convince most scientists of the bird-dinosaur connection.

Tracks in the sand tell another story, too. Spacing between prints allows for a calculation of speed. Man-sized, predatory dinosaurs could sprint at close to human speed. This, along with bone density and studies of dinosaur ecology have lead many scientists to believe at least some dinosaurs were warm-blooded—a feature present today in only birds and mammals.

If you hike long enough in canyon country, you may come across nesting great blue herons. Ten or twenty pairs of birds may come together at breeding season and build nests in trees or even on the ground. After courtship and mating, the female lays three or four pale blue eggs that hatch twenty-eight days later into some very ugly chicks. The parents don't seem to mind. They share nest duties and feed the young with regurgitated fish until they reach adult proportions.

Dinosaurs called maiasaurs that lived seventy-five million years ago in Montana, had similar parental roles. These two-ton creatures collected in colonies numbered in the hundreds, perhaps thousands; laid up to two dozen eggs per nest, and raised their hatchlings to maturity over a four-year interval. Parent maiasaurs, too, were undoubtedly kept busy foraging and regurgitating meals for their hungry offspring.

This nesting colony of maiasaurs in Montana died in the ash fall of a nearby volcano. Ten million years later, virtually all large land animals disappeared in some catastrophic event or series of events that profoundly changed the Earth's network of living things. Yesterday's successes perished, while a whole new world of opportunities beckoned small entrepreneurs. Rodent-like mammals diversified to fill the terrestrial vacancies of large dinosaurs. Small dinosaurs took to the air and their descendants, the "great blue dinosaurs," startle us with their beauty as they walk the shoreline or fish in the quiet waters of the canyon country.

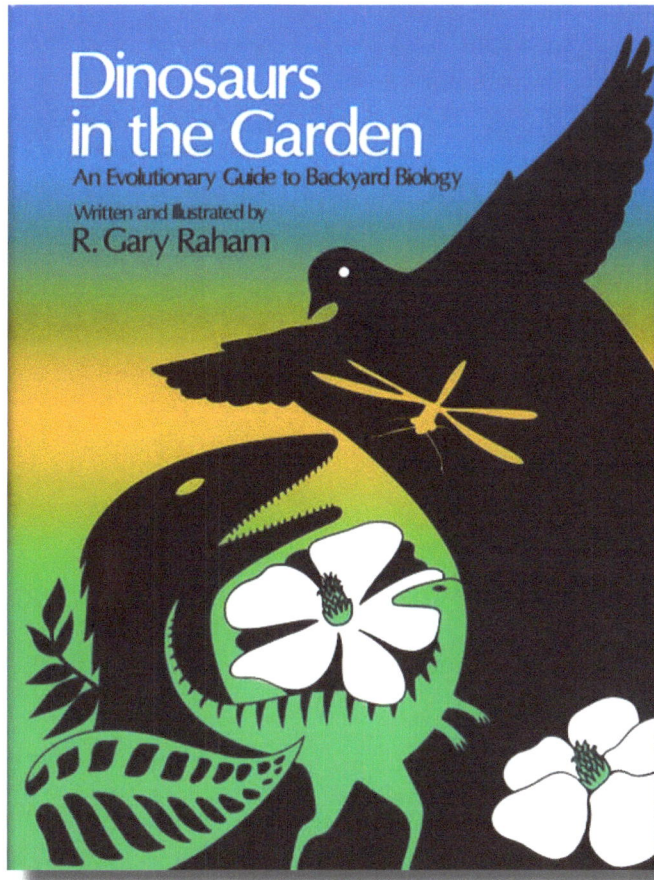

Dinosaurs in the Garden
An Evolutionary Guide to Backyard Biology
Written and Illustrated by
R. Gary Raham

In a footnote to this article, I have to mention my FIRST BOOK: Dinosaurs in the Garden, *published in 1988 by Plexus Publications. Although primarily a natural history romp through my backyard, I emphasized the long evolutionary history of each creature I featured. The book title reflected the whole "birds-are-really- dinosaurs" theme.*

I illustrated the book and designed the book cover. I was gratified when the cover art was selected to be included in the Society of Illustrators 30th Annual Exhibition in 1988. The book got some nice reviews, too, including the following:

Philip and Phylis Morrison, Scientific American, December 1988, pg. 126 – "Drawings, photographs and an informed but lively text draw out the evolutionary implications of the wonders that abound in the life around us, transmuted by insight out of the commonplace. Amusing, direct, visual, never condescending to the reader or to any other living form—this is a find."

This starling—also a clever black bird—was created on scratchboard for my first book, Dinosaurs in the Garden.

ARTICLE
Some dumb animals are pretty smart

Humans like to crow about their smarts until the keyboard smokes with conceit. But other animals (including crows) have some impressive things to squawk about, too. This article, examining the range of intelligence in other creatures, came in the wake of the discovery in August 2006 of two human RNA genes (HAR1F & HAR1R) that differ significantly from comparable genes in chimpanzees. These genes are part of a 49-gene suite called "Human Accelerated Regions" that play some role in neurodevelopment that helps make us the precocious primates that we are.

But we need to remind ourselves that intelligence still needs to prove itself over the long haul of geological time. Species, on average, tend to last about a million years before going extinct. H. sapiens has only a couple of hundred thousand years experience in the survival game and has already demonstrated that big brains can be hazardous to one's health and welfare.

This article also allowed me to show off a magpie illustration (top left) I originally created for a Natural Area sign in Fort Collins, Colorado (Magpie Meander). "Some dumb animals are pretty smart" appeared in the January 2007 issue of the North Forty News.

As humans we pride ourselves on being tops in the brains department on planet Earth. Sure, some animals can run faster, hear or smell more acutely, flash more impressive teeth and so forth, but we humans have a puffed up, convoluted lump of gray matter on the end of our spinal column that composes sonnets, builds cities, and fills out crossword puzzles. All true—but a few animals show some amazing brainpower of their own—not only our close relatives the chimps,

but also creatures as diverse—and alien—as dolphins, crows, and parrots.

Chimpanzees can learn sign language, have been observed to fish termites out of their nests with sticks—and pass on such traditions to their offspring—solve fairly sophisticated puzzles (especially when food is involved), and display complex social relationships that include an awareness of deep loss when a companion dies. In one way, it's no wonder they have such skills because they share fully 98%

of the genetic code written in our DNA. In fact, how could a 2% difference in genetic instructions make such a profound difference between KoKo the chimp and Albert Einstein?

This past August, scientists found a significant difference between humans and chimps in a nucleotide segment called the "Human Accelerated Region 1" or HAR1. This little 118 "letter" segment of DNA shows only a two-letter difference between chimpanzees and chickens, for example, which are only distantly related to each other—indicating that this segment has shown little evolutionary change in tens of millions of years at least. However, between humans and chimps, separated by only about 5 million years of evolution, 18 letters differ in the same segment. HAR1 contains two overlapping genes, HAR1F and HAR1R. Both of these genes become active during the development of the cerebral cortex—the portion of primate brains that deal with matters intellectual. Our cortex twists, turns, and expands during development much more than that of a chimp, perhaps under the direction of these genes.

But you don't have to be a primate to be smart. People have recognized the intelligence of certain whales on an anecdotal basis at least since the days of the Greek philosophers, and through detailed scientific studies on dolphins and other cetaceans since the 1960s. Like primates, the seat of intelligence for dolphins lies in the cortex of the brain, but the organization of the cortex is significantly different. For one thing, primate brains devote large parts of their neural real estate to visual sensory input, while dolphins concentrate on sound. But the dolphin cortex looks similar in terms of the complexity of its twists and turns and the ratio of brain weight to body weight is only slightly larger for humans when the dolphins extra supply of insulating fat is factored out.

Dolphins show their smarts in several ways: They engage in complex play behavior and do well on problem-solving tasks. The work of Karen Pryor in the late sixties showed that dolphins demonstrated real creativity by being able to learn that in order to get a food reward they had to perform a sequence of new behaviors of their own choosing. In a control experiment, it took humans about the same length of time to figure out the same learning task. Dolphins will also wrap pieces of sponge around their snouts to keep from bruising them—a demonstration of tool use—and they communicate with a complex mixture of clicks and whistles—including "signature whistles" that represent the names of different individuals in a social group. A study published in 2000 claims the bottle-nosed dolphin displays behavior consistent with self-awareness because they can successfully use a mirror to investigate certain marked portions of their bodies.

But it may be that crows, ravens, and other birds in the family Corvidae demonstrate the most "alien" brand of high intelligence from a human point of view. They prove that "bird brains" can be quite effective in the smarts department. Ravens, for example, can solve difficult puzzles like untangling a knotted string to free up a treat or they can steal fish by hauling in an untended fisherman's line. Crows on the island of New Caledonia have learned to use twigs to spear grubs from beneath rotting logs. In the mid 1800's, the Rev. Henry Ward Beecher once said, "If men had wings and bore black feathers, few of them would be clever enough to be crows."

Although the brain-to-body-weight ratio of crows is larger than that for dolphins and nearly matches that of humans, those brains have relatively wimpy cortical development. In the 1960s, neurologist Stanley Cobb discovered that corvid intelligence originates in a brain structure called the hyperstriatum rather than the cortex. "The larger the hyperstriatum, the better birds fare on intelligence tests," said Candice Savage in her book, *Bird Brains* (Sierra

Club Books, 1997). Crows, ravens, and magpies all fall at the high end of hyperstriatum development. Their large brains "are packed tight with exceptionally large numbers of brain cells."

Another researcher, Irene M. Pepperberg at the University of Arizona, has worked with African gray parrots—most successfully with one named Alex. Alex can identify more than a hundred objects by name, understands the concepts of "same" and "different," "absence," "quantity," and "size." Pepperberg claims that Alex has mastered tasks once thought to be the province of only humans or certain non-human primates.

So, it appears that we humans should not let our swelled heads, well—swell our heads. The science fiction writer, Douglas Adams, said in his *The Hitchhiker's Guide to the Galaxy*, "For instance, on the planet Earth, man had always assumed that he was more intelligent than dolphins because he had achieved so much—the wheel, New York, wars and so on—whilst all the dolphins had ever done was muck about in the water having a good time. But conversely, the dolphins had always believed that they were far more intelligent than man—for precisely the same reasons."

As an addendum to this article, some British researchers in 2008 Demonstrated that magpies pass another intelligence test formerly only passed by apes, elephants, and dolphins: they can recognize themselves in a mirror.

The following references may be of interest:

Heinrich, Bernd. *Ravens in Winter*. New York: Summit Books, 1989

Savage, Candice. *Bird Brains, The Intelligence of Crows, Ravens, Magpies, and Jays*. San Francisco: Sierra Club Books, 1997

R. GARY RAHAM

ARTICLE
Permafrost: Biological treasure chest & climate wild card

Some creatures demonstrate an amazing capacity for survival. Tiny creatures called tardigrades dry out under stress only to be reanimated with water decades or centuries later. Mosses can survive coverage by glaciers to survive until warmer times. Lichens hide out in crystalline rock crevices in the Antarctic, and—as outlined in this article—plant tissue has survived the last ice age to give us a glimpse of what flowered beneath the feet of mammoths and other megafauna. Such tenacity brings the pride of kinship to those of us who embrace our humble genetic connection to the rest of the biosphere.

The permafrost that allowed the survival of a strain of ice age narrow-leafed campion also represents a vast reservoir of carbon that has the potential to remake modern climate in ways we may not enjoy.

This article appeared in the November 2012 issue of the North Forty News. *The illustration is mixed media on colored matteboard and based on photographs of the resurrected plant.*

A Siberian arctic ground squirrel knew precisely where to bury her rainy day stash of seeds: in permanently frozen soil called permafrost. She never returned to retrieve the treasures in her soccer-ball sized burrow. Nearly 32,000 years later, two Russian scientists, Svetlana Yashina and David A. Gilchinsky, found her hoard and performed a near miracle: They regrew a plant that had shared the Siberian plains with mammoths, wooly rhinoceroses, and giant bison. The scientists reported their discovery in February of this year in the *Proceedings of the National Academy of Sciences.* The plant, *Silene steno-*

phylla or narrow-leafed campion, serves as an example of the rich biological material encased in permafrost, an ice age legacy totaling almost one fifth of the land area on our planet.

That same rich biological material also poses a mammoth threat, if you'll excuse the pun. A 2009 study reported by the Global Carbon Project based in Australia estimates that all the frozen plants, carcasses, and megafaunal droppings from ice ages past amount to over 1.5 trillion tons of carbon compounds—about twice the amount of carbon currently drifting in the atmosphere as carbon dioxide, methane, and other compounds. Our feverish world is melting permafrost quickly, converting megatons of potential treasure into heat-retaining greenhouse gases.

Good news first: The Russians' work shows that living things are tough (and also that human beings are pretty clever). Thirty-two thousand years in the deep freeze didn't prevent *Silene* from reproducing—although she did need a little help from her scientist discoverers. *Silene's* seeds failed to germinate, but Yashina coaxed a few of *Silene's* placental cells (like the whitish, seed-filled inner core of a bell pepper) into growing into complete and viable-seed-bearing adult plants. Placental cells, like stem cells in animals, retain an embryo-like ability to express all the genetic information in their intertwined strands of DNA.

S. stenophylla still lives today in the arctic, but the ice age variety resurrected by the Russians produces wider flower petals, slower growing roots, and more bud growth. A complete comparison of ancient and modern genomes will surely provide a roadmap to the evolution of a species adapting to a warmer, post-glacial climate.

The Russians' discovery and research provides encouragement to seed banks around the world, including the National Center for Genetic Resources Preservation (NCGRP) in Fort Collins. The Russians' work proves some seeds can successfully serve as cold-suspended templates for preserving rare and endangered plants. Dr. David Dierig, Research Leader and Location Coordinator at NCGRP said, "The important thing to remember is that not all seeds react and store the same way. There is a lot of research going on in the seed world about why some seeds are recalcitrant when other seeds (orthodox) store for very long periods of time (maybe beyond 32,000 years)."

Now the bad news: Earth's backlog of carbon trapped in permafrost over the last three million years represents a climate wild card that may tip Earth's climate into a long-term hothouse cycle. Such warm conditions—with no permanent ice at the poles—are actually more typical than our relatively chilly present. Dinosaurs, for example, evolved and came to dominate Earth ecologies during a hothouse phase lasting 160 million years.

Long-term climate patterns depend on a range of factors, including the slow drift of continents, tectonic activity, and how land mass configurations and topography influence ocean currents and atmospheric weather systems. With continents in their modern positions, celestial mechanics dictate cold cycles. Earth wobbles as it spins, its axis tilts toward and away from the sun, and its distance from the sun changes in patterns that conspire to periodically nudge the planet into ice ages. Living things also contribute to climate change by altering the levels of so-called greenhouse gases, like carbon dioxide and methane, in the atmosphere. The first forests, covering vast tracks of land in the Late Paleozoic, trapped huge stores of atmospheric carbon in their tissues, providing most of our gas, oil, and coal reserves today.

Our prolific and energy-intensive human culture has now become a global climate-changing force as we burn the remains of these long-buried forests in our cars, thus amplifying the heating powers of Earth's atmospheric greenhouse in a geologic eye blink. The National Snow and Ice Data Center in Boulder reported in September of this year that Arctic sea ice now covers just 3.41 million square miles. This is 18% lower than a record minimum set in 2007, represents the smallest area ever measured since satellite records began in 1979, and, according to *Science News* editor, Eva Emerson, may herald the end of an ice age regime that has lasted 13 million years.

The resurrection of Siberian wildflowers reveals the potential for amazing biological discoveries within the enormous, ice age sepulcher of frozen soil within the Arctic Circle. The fumes of decay from those vaults as they warm in the coming decades should also serve as a warning of the slumbering carbon elephant that threatens to move from the soil to the air and contribute to a global fever largely triggered by our own success as a species. Let's hope our instincts for survival are at least as good as those of an Arctic ground squirrel.

Though ice ages are relatively rare in Earth's history, several episodes occurred during the Carboniferous. I created this poster to illustrate that event for an "Ice Worlds" symposium for the Western Interior Paleontological Society in 2013.

Jefferson's signature fossil description was that of the claw of a giant ground sloth, Megalonyx jeffersonii—named in his honor. As explained in the text, he originally thought it to be the offensive weapon of a giant lion. Jefferson would have been flabergasted if Louis and Clark had been fortunate enough to have unearthed the fossil treasures discovered in October 2010 at Snowmass, Colorado. Among the more than 5,000 bones of ice age mammals were several examples of the claws of M. jeffersonii.

ARTICLE
Jefferson's Old Bones and the
Spirit of a New Nation

In 2001, mammoth tusks and teeth turned up during new home construction in my little town of Wellington, Colorado. That incident fit nicely with research I had done on Thomas Jefferson's interest in fossils and his desire to explore the territory acquired from the French in the Louisiana Purchase.

Jefferson possessed a keen intellect, an inquiring mind, and demonstrated considerable wisdom in crafting a document that provided the necessary structure for a new nation while remaining open-ended enough to be flexible. Such wisdom didn't always extend to decisions in his personal life, but he was merely human after all.

Jefferson was also a fellow bibliophile and compulsive reader. His personal library provided the nucleus for the Library of Congress after the British burned Washington.

Thomas Jefferson found mystery in old bones. Fortunately for all of us, Jefferson had other interests as well, including fierce beliefs in liberty, religious freedom, and the power of education. But in a way, Jefferson's fascination with fossils--and all the mysteries of nature whose secrets men were beginning to uncover--helped determine both the physical boundaries and the independent spirit of our country. The wild, uncharted western portions of North America seduced Jefferson with its secrets.

Some of those secrets would ultimately turn up in Colorado, a place Jefferson would have loved--perhaps almost as much as the wilderness of his native Virginia. He would have appreciated the sky-piercing mountains, enjoyed our wild back country, and would have eagerly examined the bones and shells that seem to pop out of every hole dug in the ground. Jefferson would have dashed to Wellington's new suburb a year ago to marvel at the mammoth tusk and molars discovered there. We look at such bones and imagine a time 10,000 years ago when herds of mammoth wandered beneath the shadow of Long's Peak. When Jefferson looked at similar bones near the turn of the seventeenth century they seemed to violate natural law. Every species on Earth was considered fixed and eternal, yet men were finding the bones of huge creatures unlike any living forms then known.

In November, 1796, Colonel John Stewart sent Jefferson three enormous claws, discovered in a cave in western Virginia. In a letter to Benjamin Rush of the Philosophical Society early the next year Jefferson said, "What are we to think of a creature whose claws were 8 inches long, when those of a lion are not 1 1/2"...?" If God's perfectly created species were truly eternal, where were the living examples of these bizarre and monstrous beasts?

Jefferson thought he knew the answer: "In the present interior of our continent," he said, "there is surely space and range enough for elephants and lions."

English expenditures on exploration in western North America worried Jefferson. He suspected they reflected English ambitions to possess whatever treasures might be there. In 1803, when he had been president for three years, Jefferson saw an opportunity to fund some exploration of his own. Using $2,500 from Congress and some of his own money, he directed Captain Meriwether Lewis to find a trade route from the Missouri to the mouth of the Columbia. Lewis enlisted his friend William Clark to share command. While they prepared to get underway, Jefferson consummated the Louisiana Purchase, giving the United States title to all the land between the Mississippi and the "Stony Mountains." The expedition produced an amazing record of the natural history and Indian cultures en route and laid the foundations of the United States Geological, Coastal, and Geographic Surveys.

Lewis found no giant elephants or lions on his famous journey to the coast, but shortly after he returned home, Jefferson sent him to supervise the removal of mammoth bones at the Big Bone Lick in Kentucky. The vertebrae of these animals were so large hunters used them as camp stools. Jefferson's excitement sparkles in a letter to Caspar Wistar, a bone

expert in Philadelphia, after the bones arrived at the White House. (The Smithsonian Institution would not be built until 1846.) On March 20, 1808, he writes: "The bones are spread in a large room, where you can work at your leisure, undisturbed by any mortal, from morning till night, taking your breakfast and dinner with us. It is a precious collection, consisting of upwards of three hundred bones, few of them of the large kinds which are already possessed."

Jefferson, in the true sense of inquiry that blossomed during the European Enlightenment and the birth of science, became a life-long student of nature's mysteries. He knew enough to know that his knowledge was limited at best, so he followed the lead of LaVoisier in Paris (the father of modern chemistry) and stuck to precise observations and measurements rather than broad speculations about new discoveries. "The moment a person forms a theory," he said, "his imagination sees, in every object, only the traits which favor that theory."

Flawed like the rest of us, Jefferson made mistakes. He spent a bit more than he earned, so he died in debt. He made some unconventional choices in matters of love. But he helped to frame a political contract, still going strong today, that allows free men to act on their own behalf in accordance with their own religious beliefs. He spent his post-presidential years creating The University of Virginia to further the education of curious farmers like himself and empower them to exercise those political freedoms.

Based on those fossil specimens sent to him by Stewart in 1796, one species bears Jefferson's name: *Megalonyx Jeffersoni*, or "great claw." It was not a huge lion as he had first thought, but a ground sloth the size of an elephant. Nor did it continue to live in the unexplored western wilderness of America, but vanished with continental glaciers 10,000 years before. Jefferson is not remembered for his contributions to science. But the sloth's weathered old bones fired Jefferson's imagination and energized his creativity in many arenas. He is remembered, justly, for helping to build a nation that "will be great in both (science and virtue) always in proportion as it is free."

The following references may be of interest:

Jefferson, Thomas. *The Writings of Thomas Jefferson*. (Andrew A. Lipscomb, Editor-in-Chief) Washington, D.C.: The Thomas Jefferson Memorial Association, 1904.

Lanham, Url. *The Bone Hunters*. New York: Dover Publications, Inc., 1973. Pp 1-7.

Osborn, Henry Fairfield. "Thomas Jefferson as a Paleontologist" in Science, Vol. 82, No. 2136, Dec. 6, 1935. Pp 533-538.

Womack, Todd. "Plentifully Charged With Fossils: The 1822 Discovery" in Fossil News, Vol. 6, No. 7, July, 2000. Pp 14-16

Smilodon *(ice age sabretoothed cat) with a Teratorn (giant vulture) in the sky (Pencil)*

ARTICLE
Beware Young Fossil Hunters:
Stay Sharp, Stay Focused and Watch Out for Mummysaurs

The story of Charles Sternberg and his sons served as the basis for two articles. The version printed here appeared as a "Kid's Quarry" feature in the October 2012 issue of Trilobite Tales, *the newsletter of the Western Interior Paleontological Society. Previously I had sold an article to* Highlights for Children *that appeared as "The Dinosaur Mummy" in their February 2000 issue. Charles Sternberg founded what might be called a dynasty of outstanding fossil hunters. His sons continued his work by hunting fossils in the American West and Canada during the 19th and early 20th centuries. Like me—and most paleontologists— the Sternbergs became captive to the lure of finding worlds lost in the catacombs of deep time.*

I didn't illustrate the Highlights *version of the story, but did create a sketch of the Sternbergs using white charcoal on blue paper for the Kid's Quarry article. The picture of the mummysaur was originally photographed by Henry Fairfield Osborn and came via Wikimedia Commons on the web.*

The artwork at the end of this article, previously unpublished, was inspired by paleontologist Edward Drinker Cope who used Sternberg as a guide and companion on some field work.

The fossil hunting bug often bites when a person is young. The passion swelled in Charles H. Sternberg as a teenage boy in the frontier state of Kansas in the 1860s. He later wrote in his autobiography, "I made up my mind what part I should play in life, and determined that whatever it might cost me in privation, danger, and solitude, I would make it my business to collect facts from the crust of the Earth..."

Sternberg followed his dream. He hunted fossils for museum displays and served as a field expert for Edward D. Cope and other scientists who became famous as dinosaur hunting academics. Sternberg did find time to marry and raise a family. This fact points to another truth: fossil hunting may be hereditary or at least contagious. All Charles' sons— George, Charlie, and Levi—became famous dinosaur and fossil hunters in their own right. What choice did they have? They discovered the bones of ancient monsters while fighting danger, loneliness, and hunger. In 1908, they uncovered something even more rare and spectacular: a dinosaur mummy.

Of course, they were looking for something else at the time.

They found it, too: *Triceratops* bones in the wilds of eastern Wyoming. Their success became part of the story. After laboring to excavate a six foot, six inch long *Triceratops* skull from its tomb, the Sternbergs were very low on supplies. Charles had to get his find shipped off to a client, the British Museum of Natural History, and stock up on food. The nearest town, Lusk, lay 65 miles away—a multi-day trip there and back by horse and wagon.

Before they could leave, the eldest son George (and a young father himself at 25) found some interesting bones sticking out of a high ridge of sandstone. It looked like it could be a major find.

Charles made a decision. He and Charlie (age 22) would take their *Triceratops* to Lusk and bring back supplies. George would stay with Levi (age 14) to protect their find and start digging around the bone bed to discover the original floor of the site and whatever lay buried there. The only problem: George and Levi would have to survive on a handful of bottom-of-the-sack potatoes and maybe a few biscuits—if they could borrow some flour from a shepherd a few miles away.

Charles and Charlie left. George found the shepherd, but he was nearly out of flour, too, and ready to leave. George and Levi would need to fuel their digging with withered potatoes.

As they dug, they began to realize they had a nearly complete animal preserved in amazing detail. That helped them ignore their growling stomachs. They exposed an area 12 feet wide by fifteen feet long, by ten feet deep. By the end of the third day, they could trace the skeleton to the breastbone. Their animal lay on its back with the end of the ribs stick-

ing up. When George removed a large slab of rock from on top of the breastbone, he stared in amazement. Twenty years later he wrote of this find. "Imagine the feeling that crept over me when I realized that here for the first time, a skeleton of a dinosaur had been discovered wrapped in its own skin."

The Sternbergs had found a dinosaur—*Edmontosaurus* (a kind of hadrosaur)—that had died in some protected spot so that it had dried out and become a natural mummy before it had become buried and mineralized. The find was purchased by and is still on display at The American Museum of Natural History. A writer later described the fossil this way: "Lying on its back, with a gaping rib cage and grinning skull…the specimen looks like a partially decomposed carcass—one can almost smell it…"

George enjoyed his father's reaction to the find when he and Charlie returned after an absence of five days. In the fading light near sunset Charles said, "George, this is a finer fossil than I have ever found. And the only thing that would have given me greater pleasure would have been to have discovered it myself."

In 1909, Charles wrote a well-received book about his experiences entitled *The Life of a Fossil Hunter.* He found many spectacular specimens, some of which reside in the Charles H. Sternberg Museum in Hays, Kansas. His sons also made paleontology their occupations. Katherine Rogers tells their stories in *The Sternberg Fossil Hunters, A Dinosaur Dynasty* (Mountain Press Publishing, 1991).

So, be forewarned: Surviving fossil hunting mania requires strength of mind and body, the mentality of a construction engineer, and often the delicate touch of a surgeon. Pay for the job is often meager, unless you are contented with the thrill of the chase and the rewards of discovering creatures—and sometimes entire worlds—never seen before.

ARTICLE
Beatrix Potter: Hare-raising Writer/Illustrator and Part Time Paleontologist

I learned snippets about Beatrix Potter's life from various sources. At some point I encountered Linda Lear's extensive biography of her: A Life in Nature. It's always sobering to find the stories of women whose natural talents and ambitions were detoured or in some cases destroyed entirely because they lived during times in which they were effectively the property of men.

Potter grew up in an affluent and relatively liberal English family, yet her destiny would normally have been to become the spouse of some (hopefully) wealthy and influential man. Fortunately, she had the determination to transform her considerable artistic talents and her interest in nature into something that would endure.

This article appeared in the September 2014 issue of Trilobite Tales, *a newsletter aimed at amateur paleontologists—specifically in Kid's Quarry, a section devoted to young fossil fanatics. Had she lived in different times her interest in fossils might have nudged her toward scientific illustration. In any case, she used her talents well as a children's storyteller and made the world take notice.*

The illustration is charcoal and pencil on gray paper.

September 1893 at Eastwood in Dunkeld, Scotland, a young woman named Beatrix Potter found and painted a rare and important fungus (*Strobilomyces*) commonly called "Old Man of the Woods." The very next day she wrote a picture letter to a five-year-old friend, Eric Moore, in which she invented a mischievous rabbit named Peter. Either of these events might have brought the young woman fame and recognition. Only one did. Today, 121 years after his "birth," Peter Rabbit stills hops into the life of many children. The world has forgotten that Beatrix Potter was also a skilled scientist and natural science illustrator. In 1893, her passions for both fungi and fossils might have taken her career in a very different direction.

Beatrix was born in 1866 to Helen and Rupert Potter. She was far from being a starving artist. Her upscale family expected her to marry and raise a family. But Beatrix was smart, talented, and interested in all kinds of things.

She wanted to do more with her life—and she did.

At the age of five, Beatrix spent her first summer holiday at Dalguise House in Dunkheld Scotland. She would spend 11 summers there exploring the woods and fields. She kept animals as pets (including a rabbit she called Peter) and loved to draw both animals and plants. She learned reading, writing, and languages from a governess. She enjoyed painting, a skill she may have inherited from an artistic paternal grandfather. Her own father became an excellent practitioner of the relatively new art of photography.

At 15 she began writing a journal in a code of her own invention. She learned to use a microscope. When she was 19 she painted her first microscopic watercolor: a gnat's leg.

By the time she reached her early twenties Beatrix's artwork came to the attention of a London publisher. She began selling work to illustrate greeting cards and books—not really for the money, but for the independence and sense of worth it provided her. She enjoyed interpreting the beauty of nature, but also sought to understand how the natural world worked.

In 1892, Beatrix met a shy old man named Charlie McIntosh—a postman by day, but also an avid student of mushrooms and other fungi. She wrote, "When one met him, a more scared startled scarecrow it would be difficult to imagine. Very tall and thin, stooping with a weak chest, one arm swinging and the walking-stick much too short, hanging to the stump [of his hand] with a loop, a long wisp of whisker blowing over either shoulder, a drip from his hat and his nose, watery eyes fixed on the puddles or anywhere, rather than the other traveller's face."

But Charlie knew his fungi and he was much impressed with Beatrix's illustrating skills. They made a great team. She gained knowledge about nature and fungi, along with specimens to draw and he acquired beautiful illustrations to accompany his scientific papers.

Beatrix drew 60 fungal specimens in the summer of 1893 at Eastwood. She also discovered the rare *Strobilomyces* mentioned above. In addition, McIntosh provided the inspiration for a character that would appear in the story she wrote to young Eric Moore the very next day: Mr. McGregor, the villain who bedevils Peter Rabbit. Potter's biographer, Linda Lear, wrote, "The ironic coincidence of Beatrix's activities at Eastwood in September 1893 would not be recognized for nearly a century. But in the space of two days she had found and painted a rare and important mycological specimen and created two fictional characters that one day would be world-famous. Both were products of her skill as a naturalist, her acute observation of people and places, her creative imagination, and her sure sense of audience."

The following year, when her family took a holiday near the village of Coldstream in Scotland, Beatrix became absorbed with fossils. She found fossil tree bark and fossil fish teeth. McIntosh advised her about how to illustrate her fossil finds. She showed the stone from various angles, how it looked both split and whole, indicating the different planes of the stone. Lear says that her natural history paintings "testify to her discipline and her desire to merge the scientific and the beautiful, revealing the truth of both."

In 1895, on holiday at Holehird at Windermere (England), she painted fossil corals found at Applethwaite beds at Sour Howes quarry. "Beatrix painted the beauty of nature unadorned" Lear said, "but her final product is always softened by the aesthetic arrangement on the page."

Beatrix made some significant discoveries, including the observation that lichens were a composite organism composed of an alga and a fungus, but this fact wouldn't be acknowledged by the scientific community until the next century. Potter even submitted a scientific paper to the Linnean Society in 1897. She submitted it under the name of an uncle (because women couldn't submit papers), but later withdrew it to do further work.

Beatrix never resubmitted the manuscript. Instead, Frederick Warne and Company took an interest in "The Tale of Peter Rabbit in Mr. McGregor's Garden." Peter Rabbit and her other charming and disarming characters became commercial successes that occupied much of Potter's time well into the twentieth century. Biographer Lear says, "Beatrix Potter had in fact created a new form of animal fable: one in which anthropomorphized animals behave always as real animals with true animal instincts and are accurately drawn by a scientific illustrator."

We can all be thankful that we have the entertaining stories of an amazing rabbit, rendered by a talented artist, but have to imagine what discoveries might have resulted if Beatrix's love of fungi and fossils had turned her career toward science.

References:

Lear, *Linda. Beatrix Potter, A Life in Nature.* New York: St. Martin's Griffin, 2007

For a chronology of Potter's life see: http://www.bpotter.com/Chronology.aspx

Amanita, *a beautiful but potentially deadly fungus (Pen & ink with marker)*

This pencil and white charcoal sketch on colored paper is based on a 1914 photograph of Knight working on a sculpture of Stegosaurus. Knight made detailed models of his subjects so that he could light them properly and make his paintings seem to be true windows to the deep past.

BOOK CHAPTER
Charles R. Knight (1874-1953)—
An Artist Who Opened Windows to the Past

Every dinosaur-obsessed child born in the twentieth century knew the images created by Charles R. Knight, although they may not have known his name. Knight, a commercial artist and wildlife painter, essentially created the modern field of paleontological illustration by taking a commission at the American Museum of Natural History and falling in love with the process of creating lost worlds and their inhabitants. As a young man, he met the famous dinosaur hunter, Edward Drinker Cope, and came under his imaginative spell.

This partial chapter, previously unpublished, is a sample of a larger work I would someday like to finish about scientist-artists who changed the perception of the world around them through their work. Artists with the critical skills of trained observers often make important discoveries in their own right that they can successfully translate into educationally effective and strikingly beautiful images.

The opening scene is only partly fiction, drawing heavily on Knight's own words in his Autobiography of an Artist, unfinished and unpublished until 2005 by his daughter, Rhoda Knight Kalt.

March, 1897, 2102 Pine Street in Old Philadelphia

Charles R. Knight knocked on the door of the small, three-story red brick house. He straightened his tie with one hand and gripped the edges of his portfolio a little tighter with the other. He looked forward to finally meeting Dr. Cope, a legendary genius in the field of paleontology. The door opened. One of Dr. Cope's assistants greeted him warmly and ushered him inside. Dr. Osborn had confided to Charles that Cope's fortunes had taken a turn for the worse of late, so Charles wasn't quite sure what to expect.

Charles adjusted his thick glasses and glanced around. The house stood in a jumbled chaos worthy of a Dickens' novel. Dust lay everywhere; piles of pamphlets formed ragged pillars. To Charles' right, in the front parlor, the massive bones of some extinct dinosaurian monster littered the floor. The windows were shuttered and the walls bare of curtains or pictures of any kind.

The assistant led Charles up narrow stairs

bordered with more pamphlets stacked floor to ceiling. Pickled snakes and reptiles in glass bottles stared back at him from shelves along the walls. The stairs led to a sitting room on the second floor. A bay window overlooked a meager garden. Dr. Cope sat behind a desk piled high with books. A human skull grinned from the mantle behind him and a large bronze vulture spread its menacing pinions above a cage with a live Gila monster. Someday Charles promised himself that he would sketch this scene.

The doctor smiled and offered his hand. "Mister Knight, welcome. I have heard good things about thee from Dr. Osborn." The paleontologist's Quaker speech pattern sounded quaint to Charles' ear.

So began a two-week visit with the great scientist where Charles spent part of the time sketching at a drawing table by the bay window (the only free space in the room) and part of the time listening with rapt attention to the greatest conversationalist that ever graced the service of paleontology. Under the spell of Cope's facile tongue new vistas of the life of the past opened before Charles in a way he had never dreamed of…

A sketch of Edward Drinker Cope in his prime

The scene above, based largely on a passage from Charles R. Knight's autobiography, describes the first meeting between the artist, at age 23, and the famous fossil hunter, Edward Drinker Cope. Henry Fairfield Osborn, another scientist destined for fame and a lifelong friendship with Knight, had introduced the pair. Both scientists would inspire this young wildlife painter to attempt something entirely new: the recreation in art of worlds long past using clues from the fossilized bones and other remnants of its inhabitants. Knight went on to inspire legions of 20th century artists, movie-

makers, and writers by lighting their imaginations with vistas filled with strange lumbering beasts and haunting landscapes. His images helped make names like *T. rex, Triceratops*, and *Velociraptor* tumble from the lips of every wide-eyed child.

Knight's career, spanning seven decades, began with a simple love of animals. According to Knight's father, the first complete sentence he uttered at the age of two was "See the black chicken!" At age five, Charles spent hours copying animal illustrations from books that he loved. Somewhat later in life, he thought it ironic that he was a wildlife artist born in the big city of Brooklyn, New York, where he had to haunt Central Park Zoo to see and draw wild animals on a regular basis. There was also tragic irony in the fact that for a person destined to depict the ghosts of dinosaurs and other prehistoric beasts in vivid detail, he navigated his day-to-day world in the misty veils of severe astigmatism. He was legally blind.

Knight inherited near-sightedness from his father, but he suffered another blow (both literally and emotionally) at age six when a young boy tossed a pebble that struck him in the right eye. A doctor prescribed bed rest in a darkened room. Knight said in his unfinished autobiography (published by his granddaughter in 2005), "Here for six weeks I lay in a sort of misery, very little pain, but most uncomfortable, while dressings of acetate (sugar) of lead were kept continually over my injured optic."

Though it seemed that he had recovered completely, the incident caused lasting damage to the cornea that eventually showed up as an inability to see things clearly at a distance. One of Knight's greatest fears as he grew older was how the increasing loss of sight would impact his work as an artist.

Early inspirations

Knight felt destined for a career in the arts, but several things conspired to put him on the path of science-inspired illustration. His father, George Wakefield Knight, shared his love of fishing and hiking with his son. The elder Knight served as secretary to the wealthy financier, James Pierpont Morgan (1837-1913). J. P. Morgan, in addition to being president of the Metropolitan Museum of Art, donated heavily to the American Museum of Natural History (AMNH). George Knight treated his son to many weekend museum visits. Knight said, "The long tiled halls filled to overflowing with glassy-eyed birds and animals, each on its own mahogany base, fascinated me and we had them all to ourselves, as the museum was not then open to the public on Sundays."

Not long after his eye injury, Knight also

lost his mother to a brief battle with pneumonia. In 1882, his father, G. W. Knight, married Sarah Davis, a painter. Sarah provided Charles with the template of a practicing artist. She encouraged her stepson to pursue art education at various New York schools, including the Froebel Academy and the Metropolitan Art School. Nevertheless, mother and son butted wills on many occasions. Sarah may have been a little jealous of Charles' precocious talent. In 1890 at age16, Charles left home and found a job with J & R Lamb creating animal images for stained glass windows.

Just two years later, the elder Knight died. Devastated, Charles returned home for a year, but then moved to Manhattan to begin a freelance career as a children's illustrator. He also illustrated extensively for *McClure's* magazine. He still visited zoos to sketch and spent hours in the taxidermy department of AMNH. It was at the latter location one day when a friend referred him to a Dr. Wortman who was looking for an artist to reconstruct an extinct pig-like animal called *Elotherium*. That referral changed the course of Knight's professional life.

Knight created an excellent reconstruction of *Elotherium* that obtained him further work from Dr. Wortman. A young paleontologist named Henry Fairfield Osborn also noticed Knight's work. Osborn had plans to revolutionize the way museums displayed and interpreted their fossils and he saw Knight as someone who could help make that vision come true. As Knight said in his autobiography, "Thus in the most prosaic way imaginable I was introduced to a set of men whose interest and encouragement eventually opened up for me a momentous period in my life work, and created a whole new field of research and study into which I could delve to my heart's content…"

Knight's personal timeline and career highlights

Knight and Osborn shared the vision of

bringing the past to life, but their differing perspectives and temperaments generated some friction. Knight felt secure in his artistic skills and wanted to maintain control over the esthetic aspects of his paintings. Osborn needed to maintain scientific accuracy and would have loved to make Knight an employee of the museum—something Knight didn't want. Knight preferred to remain an independent contractor, but always struggled with financial details that he found tiresome, however necessary they might be.

Knight continued to work for *McClure's*. In 1896, the owner, S. S. McClure, sent him on a European tour that provided an opportunity for Knight to sketch animals from zoos all across the continent and to meet other animal artists. His association with the magazine introduced him to writers like Rudyard Kipling and Sir Arthur Conan Doyle.

Back home at AMNH, Osborn introduced Knight to the famous fossil hunter, Edward Drinker Cope—an event dramatized at the beginning of this chapter. Cope was a talented amateur artist who illustrated many of his own manuscripts. Osborn later wrote Cope's biography, *Cope: Master Naturalist* (1931). Knight learned much about Cope's work and life experience during the fossil hunter's last two weeks of life. He also absorbed some of the scientist's intense passion for bringing the past to life.

In 1900, Knight married Annie Humphrey Hardcastle. She was a strong-willed Southern woman, according to granddaughter Rhoda Knight Kalt, who complemented her "dreamy, impractical artist" husband. Annie—and later their daughter, Lucy, kept Knight's business life and finances in order.

During the first decade of the twentieth century, Knight created many paintings for AMNH, but also produced work for the U.S. Fish Commission and the U.S. Bureau of Fisheries. Knight's prehistoric reconstructions became the biggest crowd pleasers at the museum and Osborn and Knight planned for an ambitious undertaking that would illustrate the entire grand panorama of life on Earth. In one letter to Knight Osborn wrote, "I have my heart set on this great series of halls, unsurpassed in any part of the world, being illustrated by your masterly hand under my scientific and artistic direction."

Sadly, funding never materialized for the big endeavor.

In the years from 1910 to 1923 Knight completed a series of murals for the museum's Hall of Man exhibit. J. P. Morgan also set him up with a quiet studio off the museum grounds where he could work without interruption. In 1925, he painted a mural illustrating the famous LaBrea Tar Pits for the Natural History Museum of Los Angeles County, California.

But in 1926, with the help of his business-savvy daughter, Lucy, he landed the biggest contract of his career with the Field Museum of Chicago to create 28 murals illustrating the history of life—essentially the same project that he and Osborn had dreamed about. Generations of students—including this author in the 1960s—have, and continue to visit the prehistoric time exhibit, falling under the spell of Knight's visions.

During the time of his work in Chicago in 1927, he took another European trip to see the famous prehistoric cave paintings in France and Spain. The scientist Abbe Breuil served as his informative guide. "He was a man of great cultivation, a keen and mischievous host, and a veritable mine of knowledge on many subjects." This trip inspired illustrations that later ended up in one of the books he wrote: *Prehistoric Man: The Great Adventure* (1949).

After completing the murals for the Chicago museum, Knight continued to produce work for AMNH. He also wrote for *National Geographic*. In 1935, he authored and illustrated his first book, *Before the Dawn of History*. In November of that same year, his good friend and collaborator, Osborn, passed away. Knight wrote the following poem:

How peacefully he sleeps!
Yet may his ever-questing spirit, freed at length
From all the frettings of this little world,
Wander at will among the uncharted stars…
And beasts and men that to his earthly sight
Were merely bits of stone
Shall live again to gladden those eager eyes…

Knight wrote and illustrated *Life Through the Ages* in 1946 along with a volume for artists in 1947: *Animal Drawing*. Like many creative individuals, Knight worked until near the end of his life. Because he had to get closer and closer to his work as his vision failed, he created finished paintings at smaller size that he then projected onto the final surface. Assistants painted large portions on the wall and then Knight could attend to smaller detail work. He painted his last mural in 1951 and passed away peacefully at the Polyclinic Hospital in Manhattan, April 15, 1953. The last thing he said to his daughter Lucy was, "Don't let anything happen to my drawings."

She didn't.

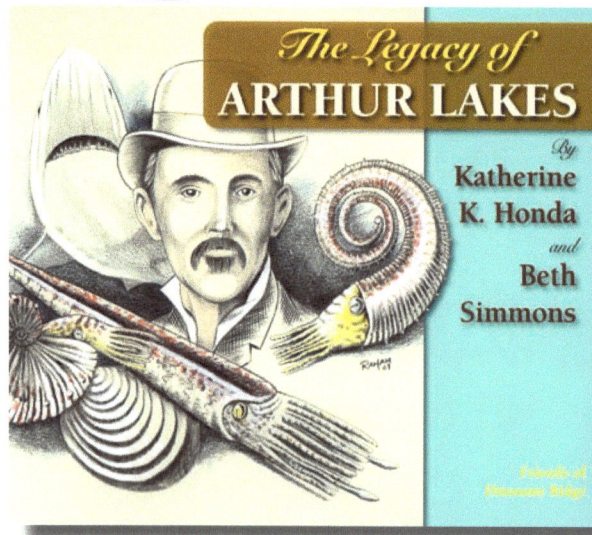

The Legacy of
ARTHUR LAKES

By
**Katherine
K. Honda**
and
**Beth
Simmons**

ARTICLE
Arthur Lakes and the sea monsters of Fossil Creek

It's gratifying to live in one geographical area long enough to acquire an intimate sense of place. That rather vague term for me encompasses not merely terrain, but the accumulated bio-geophysical history that molds and infuses character into every locale. I knew I had developed that sense of place for Northern Colorado as I studied the life and times of Arthur Lakes—partly as a result of designing a book for local geologist, Dr. Beth Simmons, who adopted Lakes, dug into his legacies and co-wrote a book with Katherine Honda detailing his Colorado exploits. I felt a certain kinship with Lakes that grew out of similar preoccupations with ancient life, the shared skills of an artist, and a common love of a rugged, but scenic land.

I donned a bowler hat and temporarily became Lakes for a school group in Fort Collins, showing them where he and his students jumped off the train six miles south of the fledgling Colorado State University campus. With a rudimentary English accent I described in florid prose—as he was alleged to have done for his students— how an ancient seaway had shaped the land 70 million years before.

My illustration of Lakes eventually became the cover of The Legacy of Arthur Lakes *and several years later I presented the original art to Beth at a Dinosaur Ridge fundraising event at Red Rocks amphitheater—a long stone's throw from where Lakes discovered Colorado's first dinosaur remains.*

The article appeared in the North Forty News *in 2007.*

Arthur Lakes preceded me in Colorado by 100 years, but I feel we would have made great friends. We both practiced teaching and art; we both became captivated by a sea level version of Colorado whose former existence sometimes asserts itself when beautiful and somewhat mysterious fossil shells and sea monsters erode from clay soils and sandstone ridges. I will get a chance to reenact a moment in Lakes' life this October 21st, just 129 years and three days after he jumped off a slow-moving train with some of his eager students where railroad tracks cross fossil creek in what is now Red Tail Grove Natural Area just south of Harmony Road. This will be "Ramble 11" in a series of field trips hosted by The Friends of Dinosaur Ridge, an organization that educates the public about the dinosaur footprints and fossil remains still visible in the hogbacks near Morrison, west of Denver. (http://www.dinoridge.org)

1878 was a tough year for Lakes. The

school at which he had taught for nine years burned to the ground and he was out of work.

He had made a life for himself teaching at an Episcopal Boy's Prep school in Golden, Colorado, after attending Queens College, Oxford England, and immigrating to the United States. He began teaching in 1869, and became a deacon in the church and itinerant minister for Idaho Springs in 1874. He most likely would never have entered history books at all—although he did go on to become a distinguished professor in the earth sciences—if he hadn't discovered the first dinosaurs in Colorado and got himself stuck in the middle of a feud between two rich academics of the day, Edward Drinker Cope and Othniel Marsh—but that's another, if somewhat related story.

Lakes had just discovered the bones of *Apatosaurus*, a giant sauropod, and *Stegosaurus*, another jumbo veggie-eating dinosaur, in the spring of 1877. Marsh, eager to find bones of these then exotic giants, offered to pay Lakes to find some more—at least for a few months. Marsh invited Lakes to New Haven, Connecticut, to see his academic collections at Yale and seal the deal. Between April when the school

burned, however, and his trip back east, Lakes and a party of students had some time to explore the fossil-rich, semi wilderness just 6 miles south of the nearly completed agricultural college at Fort Collins.

According to the State School of Mines' *Colorado Transcript* of Oct. 23, 1878, "The weather was propitious, and our party, but for the presence of ladies and the Rev. J.R. Eads with us, might have been mistaken for a squad of navies or coal miners on their way to Erie armed with heavy sledges and picks." The travelers enjoyed some spectacular scenery along the way and were treated to professor Lakes' vivid tales of extinct creatures. The *Transcript* said that with a little imagination a cow on the track might easily be transformed into a giant *brontotherium* (Think huge rhino with a multihorned snout.) out for a stroll.

The scenery included a wall of basaltic lava near Boulder called Valmont Butte and "the magnificent mountain mass of Longs Peak (14,370 feet), the summit of which is cloven from top to timberline by a sheer smooth perpendicular cliff 3,000 feet high." The scientists also observed that "the little towns of Berthoud

View of a train on the same tracks used by Arthur Lakes in 1877. Photo taken by the author in October, 2007 with Long's Peak in the background.

and Loveland seemed to be building up fast, although but of a year's growth."

But excitement rose as the party neared Fossil Creek "for the conductor told us when the train slackened up we must get ready to jump for they could not stop a minute." And so they jumped, while the train moved on to the north, and they found themselves in what almost seemed like a battlefield "with the ground literally covered with what appeared both from form and color to be cannon balls and bomb shells." Instead, they looked upon concretions—balls of sediment, many of which concealed the mortal remains of relatively familiar, if over-sized clams and oysters, while others held the coiled or cone-shaped shells of squid-like animals long extinct. One coiled, ammonite shell found on the trip measured 28" in diameter.

Professor Lakes was now in his element. He told his students "You are sitting like mermaids and mermen on the bottom of a primeval, tropical ocean, formed by an arm of the sea which extended from the Gulf of Mexico along the base of these mountains to the Arctic regions. Proofs of its tropical character are in the corals and shells which you have been gathering, which only live and grow in tropical waters."

The author of the *Transcript* article marveled at the depth and extent of this vast ocean. "How long did it take 7,000 feet of little microscopic insects to live and die and form solid rocks? How long? The world is very old!"

The party spent a long and satisfying day hunting fossils and walking the six miles into Fort Collins to catch a train ride back south. Lakes worked for Marsh for a few years and made significant discoveries at Como Bluffs, Wyoming, many of which Marsh took complete credit for—a not uncommon practice at the time. Lakes was, essentially a field man. In 1880 Lakes taught geology and mining at the Colorado School of Mines until he resigned in 1893 to work as a mining magazine editor. He and his sons later became mining consultants. In 1913 he moved to Canada to be closer to family there. That's another interesting synchrony in our lives: my roots go back to England by way of Canada as well.

Lakes jaw would have dropped to see how the rugged Front Range he traveled with some difficulty has been populated and tamed, but the trains still pass down the tracks in the shadow of Longs Peak, and sunlight still outlines the concentric rings of shells trapped in the turned earth of subdivisions and Wal-Mart parking lots.

The ancient seaway abides.

Many thanks to W.I.P.S. member, Beth Simmons, for sharing research information uncovered for her biography of Lakes entitled *The Legacy of Arthur Lakes*. (Friends of Dinosaur Ridge, 2009)

Discovering Dinosaurs in the West, The Field Journals of Arthur Lakes (Smithsonian Institution Press, 1997), edited by Michael Kohl and John McIntosh also provides a good read.

I created this scratchboard image of Eiseley as he looked in the 1920s specifically for this article. Some people have told me I look a little like him. It may just be that bespectacled academic ambience.

ARTICLE
Loren Eiseley, haunted by the ghosts of Lindenmeier

Loren Eiseley (1907-1977) belonged to my parent's generation, but he spoke to me about the mysteries of nature and deep time with a dark, almost Poe-like elegance I could only admire. He spent a lonely childhood with a deaf mother and an often-absent father trapped by the constraints of the Great Depression. Eiseley often hid out in libraries letting words transport him to times and places glazed with the sense of wonder and possibilities that he hungered for.

He excelled as a poet and writer focused on nature. As a senior in high school in 1924, he wrote a short story that received an honorable mention in a contest sponsored by the Atlantic Monthy. He pursued an academic career at the University of Nebraska, but because of limited finances and some struggles with TB, it took him eight years to complete his degree. A scholarship took him to the University of Pennsylvania. He studied science, probably because it offered more career possibilities than writing, but he considered his pursuit of science to be a kind of camouflage. Loren received his PhD in 1937, but his biographer, Gale E. Christianson said, "The poet walked off the stage—diploma in hand—wearing the fox skin of a scientist."

Nevertheless, as a scientist he made one discovery that forever changed our perception of early man in North America.

This article first appeared in the June 2009 issue of The North Forty News.

I can't recall exactly when I encountered the writing of Loren Eiseley, but he conjured the past with sentences so elegantly crafted that I immediately became a fan. And he was more than just a wordsmith. As a practicing anthropologist, he actually made critical discoveries that expanded our knowledge of people and cultures that dissolved with the melt water of the last ice age. When I moved to the Fort Collins area in the 1970s I was delighted to learn that his name was linked to the discovery of a finely crafted Folsom point embedded in the vertebra of an extinct giant bison—graphic proof that, over 12,000 years ago, artisan-hunters of the first caliber felled giant animals where we now park our Subarus and build Walmarts.

In 1924, district court judge Claude C. Coffin and his son found unusually crafted arrowheads at a location that would eventually be named after the owner of the property, William Lindenmeier, Jr. The thin, tapered points made from jaspers and chalcedonies sported delicately fluted edges and variously shaped bases framing a grooved central blade. They might have been called Lindenmeier Points if the discoverers of similar artifacts near Folsom, New Mexico, hadn't published their findings first.

Claude and his brother Roy, a geologist at the then Colorado State College, collected and described material for several years. In 1930 Dr. E. B. Renaud of the Department of Anthropology of Denver University declared that the artifacts found by the Coffins closely resembled the New Mexico Folsom points. Between 1935 and 1940 the Smithsonian Museum was invited in to work the site. Eiseley made his historic discovery in 1935.

Eiseley believed in what he did, but with reservations. In his autobiography, *All the Strange Hours*, he said, "Men should discover their past. I admit to this. It has been my profession. Only so can we learn our limitations and come in time to suffer life with compassion." But he went on to say, "I now believe that there are occasions when the earth tells our story just as well, when the tomb should remain hidden, the dead man masked in jade be allowed to lie sleeping at the temple's heart."

Eiseley, the son of a deaf mother and salesman/part-time Shakespearian actor father, tumbled into his academic career in an unusual way. He acquired TB as a young adult and spent time improving his health while caretaking some property near Death Valley, California. He had little money and spent at least a few weeks as a hobo on freight trains that traversed the American West. Finally, in 1925, he decided he wanted to go to college, but had to pursue that ambition in fits and starts. He finally acquired the credits for an undergraduate degree in 1933. His academic prowess earned Eiseley a scholarship from the University of

Pennsylvania and there he met a kindred spirit in teacher Frank Speck. "In return (for helping you) he wanted very little," Eiseley said of Speck, "an exclamation over a rare fern, something in the way of beliefs shared as though by two men who sat before a brush shelter in the flickering dark of a campfire…"

So, in 1935, he found himself hunched over the remains of an ancient paleo Indian kill site, admiring the skill of someone who "had loved his instrument and so embellished it/ that a man centuries away would finger the stone lovingly,/ momentarily forgetting its purpose/ in the greater glory of art –"

In 1972, in a collection of poems called *Notes of an Alchemist*, Eiseley expressed his reservations about the Lindenmeier discovery. In a poem titled "Flight 857," he remembers the discovery as he is flying into Denver during a blizzard. He says,

> "I know now
> it should never have been resurrected
> any more than these wheels and wings and
> electronic voices
> should ever again be lifted
> from oblivion.
> I hope they do not find us:
> The point should remain in the vertebra,
> the offering by the dead child in the cave,
> the pterodactyl in the slate,
> the poet in the lost book,
> the singer as song in the grass.
> Why must we usurp
> the autumn leaf's prerogative
> or the cancellations of running water
> or the erasures of the dust?
> Like the hunters, we will leave deadly slivers of glass
> where they left flint,
> the metal will oxidize.
> We will be dangerous if found
> by anything wiser
> than a field mouse."

The Lindenmeier site now resides on the Soapstone Natural Area that opened to the public June 6, 2009. The Smithsonian excavation has been long reburied, but the City of Fort Collins has placed a kiosk overlooking the area.

Some of Eiseley's popular titles include *The Immense Journey* (1957), *The Firmament of Time* (1960), and *The Unexpected Universe* (1969). Gale E. Christianson, professor of history at Indiana State University, discusses Eiseley's life in some detail in his biography, *Fox at the Wood's Edge* (1990).

I sketched this image in pencil on paper with a pebbled surface shortly after reading Eiseley's autobiography. The imagery reflects Eiseley's preoccupation with lost worlds, dark places, and the serendipitous pranks of time and evolution.

Kaiah: Pencil sketch on matteboard

Part IV:
Confessing to creative fictions

ARTICLE
Visit to a challenging future
Crystal Lakes, Colorado, October 2049

In 2007, I wrote the following science fiction vignette for the North Forty News *projecting what Colorado might look like in the year 2049. To personalize it for myself, I imagined my granddaughter, then 8 years old, grown up and visiting property that my wife and I bought at Crystal Lakes in the late 1970s. I then "translated" my fictional speculations in terms of technological and climatic trends evident in 2007.*

Not surprisingly, my crystal ball now seems a bit fuzzy in some regards for certain prognostications. Seven years ago I talked about global atmospheric carbon dioxide levels being at 380 part per million (ppm) and that it would be prudent to limit such levels to 400 ppm by the year 2100 as levels of 500 to 600 ppm 40 million years ago were associated with tropical forests in Wyoming. As of November 5, 2014 average atmospheric levels were reported to be 395.93 ppm at CO2now.org. The web site also noted that levels temporarily reached 400 ppm on May 10 of 2013.

As a student of the deep past, I may worry more than most about the rates of change we are creating on the planet and what those changes may mean for both my descendants and yours. Hopefully, we will be wise enough to heed the signs of our impacts and clever enough to preserve a rich and beautiful natural world for those who come after us.

"Hey, Mom, look—a meteor," said Kent. His raised arm glowed in the flickering firelight.

"What a great birthday present!" Kaiah smiled at her teen son and daughter, Kay Celeste. *We certainly wouldn't see that anywhere near Denver,* she thought to herself, imagining the seemingly infinite pinpricks of lights that speckled Colorado's entire front range from border to border. Their glow diluted the stars, essentially erasing their majesty from the thoughts of most of the 7 million people living in the state.

"Great idea to come up to Crystal," Kent said, as he took a bite from his nearly blackened marshmallow. "When will Dad get here?"

"I heard from him just a few minutes ago," Kaiah said. He's on the new Maglev and is going to rent a SmartzCar in Wellington. That way he can get some work done until the PubTrans strip runs out at The Corners.

"Toast another 'mallow with us Mom," said KC. "Just make sure you save some for Dad."

Kaiah smiled. The MedAlert on her wrist glowed a reassuring green. Her sugar levels must be fine. She decided another marshmallow sounded *just perfect. She settled down between her two teens and lowered a marshmallow near the glowing coals, sighing as she found a comfortable spot on the rocks.*

"We going to have enough carbon credits to go to Greenland this summer?" asked Kent, trying to lick a glob of white from his upper lip. "I hear the skiing is going to be great!"

Kaiah remembered when Colorado was the place to be if you were a skier, although even when she was a kid the ski season kept getting shorter and shorter. "Dad and I've got it covered," she said to Kent.

"Cost us a summer at the reservoir," said KC. "I wanted to invite Theo."

"Reservoir's too low anyway, honey," said Kaiah. "The docks are high and dry and there aren't enough construction mechs to do the refitting before next year. I heard they converted a lot of them to help fight the Vail fire last year."

Kaiah looked at Kent with his marshmallow and saw the boy again in her young man's eyes. She hoped he would stay the course at University and not get drafted. With both the Mid East "Water Wars" in full swing and the military action in the Dead Zone after the New Delhi catastrophe, he was bound to end up some place that would give her more gray hair for sure.

Fifty is a good age for reflection, Kaiah mused. She'd had a good life until now and expected many good years yet to come, but wondered what it would have been like to grow up in a cooler, less crowded world like her parents—and especially her grandparents—a world when you didn't have to worry so much about basic things like fresh water, fresh air, room to live—and room to wonder.

Kaiah leaned back against the Ponderosa behind her and enjoyed the moment. A bat's wing flickered past the moon and there—right there near the belt of Orion—another meteor flashed and died across the black, northern Colorado night.

Even science fiction writers find their crystal balls murky and imprecise, but the effort is worthwhile. A meander through the fictional exercise above will demonstrate what powers this particular glimpse of a possible northern Colorado future.

7 million people in Colorado by 2049?

This estimate comes from a publication called "Colorado's Population in 2050, A Road Paved with Good Intentions," by Leon F. Bouvier and Sharon McCloe Stein. Colorado's population grew from 1.3 million in 1950 to 4.3 million in 2000. Over this period, 38% of the growth was from Coloradoans making more Coloradoans, over 50% was from net immigration from other states, and the remaining 12% or so was immigration from foreign countries. Assuming these trends continue, we can expect around 6.4 million people in 2025 and reach the 7 million person milestone by 2050.

Maglevs, SmartzCars, and PubTrans strips

Maglev stands for "Magnetic Levitation"—a wheel-less mass transit technology envisioned in the U.S. in the 70's, but first commercially built in Shanghai in 2002 with German engineering. Whether maglevs are actually built along Colorado's front range corridor or something more traditional, some sort of mass transit will make sense in an energy-starved world and could use the existing railroad right of way—especially if the railroad moves its tracks east, as some have speculated. Cars are already talking to us and refusing to let us drive if we're drunk. In the late 1990s, prototype systems for "driverless cars" were built in Italy and the United States. Cars automatically steer themselves by sensing painted lines or magnetic monorails embedded in the road.

MedAlerts, personalized medicine, and the transparent society

The growth of the biological sciences,

especially when it comes to understanding in more detail how the human genome operates and varies from individual to individual, will most likely result in automated ways to monitor some of those medical conditions for which we are genetically predisposed.

Another prospect, sincerely scary to those of us who grew up reading George Orwell's 1984, is a society that protects itself by insuring that no one has any privacy. This so-called "transparent society," proposed in a nonfiction book by SF writer David Brin and fictionalized in his book, *Kiln People*, would place surveillance cameras everywhere and provide anyone access to communication devices and databases so that no single privileged "Big Brother" could run the table.

Global warming and a world in climatic flux

The Earth is heating up (even though we are still officially living within a warm interlude of one of her most recent "Ice Ages.") The evidence has been graphed, recorded, and plotted for some time now and compared with the periodic rise and fall of prehistoric temperatures, as recorded indirectly by several corroborating techniques. The rapidly rising spike in the greenhouse gas, carbon dioxide, recorded continuously since 1958 by sensors on the Hawaian volcano Mauna Loa, parallels human population growth and technological prowess. Globally, CO_2 has risen from 280 parts per million (ppm) since 1850 to 380 ppm today. By 2100, under a "business-as-usual" approach, CO_2 could conceivably reach concentrations not seen for 40 million years—500 to 600 ppm—a time that featured little or no ice, even at Earth's poles. Paleontologist Peter Ward in his 2007 book, *Under a Green Sky,* warns that such a warm world would not be a human paradise. In fact, he asserts that most of the major extinction events during Earth's prehistory—periods of time during which up to half of ALL living species disappeared—can now be associated with high CO_2 concentrations and runaway global warming.

In the past, such high greenhouse gas levels have been associated with long and intense periods of volcanic activity that far exceed anything experienced during all of human history. Now, human activities appear to be a prime cause. Some even argue that the development of agriculture some 10,000 years ago began the trend and has helped to delay the pulse of cold that has normally followed brief warm periods over the past million years of episodic glaciations.

An organization called The Rocky Mountain Climate Organization based in Louisville, Colorado, provides lots of helpful information about the potential for local climate warming on their website at www.rockymountainclimate.org. Businesses and municipalities have begun planning for this change, taking a leadership role that some levels of government have ignored or shunned.

And while warming will occur, human action can ameliorate the trend and minimize some of the consequences. Ward believes if we begin to make sharp cuts now in carbon dioxide emissions, we could limit concentrations of greenhouse gases to around 400 ppm by the end of the 21st century. Humans have been carrying out an uncontrolled experiment with the Earth's climate for centuries if not millennia. Ward quotes climatologist Wally Broecker who said "The climate is like a wild beast, and we're poking it with sticks." Ward emphasizes that counter action to reduce global warming during the next fifty years will be critical in blunting impacts that will surely occur as sea levels rise, ocean currents shift, and their chemistries slowly alter from oxygen rich to oxygen poor and acidic. Rising sea levels will displace millions of humans and drown prime river delta farmland. Shifting ocean currents could make Europe colder while other parts of the world warm. Changes in ocean chemistry could ultimately poison the atmosphere with gases like hydrogen sulfide produced by oxygen phobic microbes.

By 2049 we can expect an average worldwide temperature increase of somewhere around 1.5^0 F (with a range between 0.5^0 F and nearly 3^0 F). In Colorado we can expect to see more episodes of drought and more extremes in weather as added heat energy stirs the atmospheric pot. This increased heat will lead to several consequences in Colorado: Less snow pack, less water overall, and more wildfires. These trends, in turn, will be exacerbated by population growth and impact Colorado's sources of income.

The good news is that a rich and meaningful future for our grandchildren and their children is not an impossible dream, but it will require three things:

1.) Recognition of the problems generated by global warming and a growing human population

2.) A political willingness to act to insure that changes are made in the way we do business and live our lives

3.) A view toward the future that takes heed of Earth's deep time past and some of the stresses that changing atmospheric and oceanic chemistry have placed on living things.

Fulfilling these requirements will require imagination to visualize the future we want and the dedication and will to make it so. Some of that imagination and dedication will come from reconnecting with nature and recognizing our place within it—somewhere between a Ponderosa and a shooting star.

Photo montage used in original article. The green hat was Grandma's, often used on mountain hikes during her college years

Kaiah and Grandma on top of the world at crystal Lakes.

The Dinosaurs' Last Seashore

And the creatures that transformed a world

Written & illustrated by

Gary Raham

GRAPHIC SHORT STORY EXCERPT
The Dinosaurs' Last Seashore

As I've said in earlier articles, writer Loren Eiseley influenced much of my work. The Dinosaurs' Last Seashore was my attempt at a graphic short story based on one of Eiseley's early essays: "How Flowers Changed the World," which I encountered in his 1957 book, The Immense Journey.

Again, I indulged in a paleontologist's fantasy—to actually wander the distant shores of a place and time too ancient to truly comprehend. I attempted to use language I hoped would mirror Eiseley's prose and illustrated my text with mostly black and white pen and ink or scratchboard renderings. I painted the full color cover art before I wrote this book to depict Baculite Mesa, a paleo hunting ground for Cretaceous fossils I visited on several occasions. Malcolm Bedell, a Western Interior Paleontological society member led these expeditions. His wife, Susan Passmore, gifted him with the original art.

I published The Dinosaurs' Last Seashore in 2010. It became a finalist for the Colorado Authors' League Awards the following year. I've excerpted the beginning to provide a setup and a sampling of an interior spread.

110 million years ago the ocean invaded what is now North America. Water ebbed and flowed for 45 million years as continents slowly ground their way into modern configurations. On the beaches I visit, near the end of the dinosaurs' reign, days are mostly warm but stroked with mild breezes. Tall evergreens create a watchful wall of green along the shore. Pteranodons soar overhead, quiet until migrating herds of hadrosaurs or sauropods bellow beneath them, following sandy highways from breeding grounds to feeding fields and back again. Time abides.

I love coming to this place and time: The Great Western Interior Seaway of North America 67 million years ago. I admire the pterosaurs that posture and groom themselves on the barrier islands and the long, graceful necks of distant plesiosaurs. Humid mangrove swamps fill the lowlands, each tree heavy with hanging vines. Vast rolling plains farther inland host herds of dinosaurs that seem to crawl across an infinite green carpet littered with islands of forest. The cliffs, dusted with pines and gaudy ferns, turn burnished gold in the evening sunlight.

As permanent as this world seems, it will end within two million years—a geological instant enfolding eternity for individual species. An asteroid will bring havoc to the planet, but that hunk of stone will only accelerate a revolution already underway—one caused by creatures you might least suspect. They will set the stage for mammals and, incidentally, for us--*Homo sapiens*--the "wise" ones. I want to see these benefactors firsthand, giant slayers sewn into the fabric of an ancient paradise. If that quest appeals to you, join me. I'd love the company.

As we travel, enjoy the strange, enormous, and sometimes savage beasts that call this place home.

When and where I come from, the oceans are gone and some of these beautiful and fearsome creatures have piled up one on top of the other and turned to stone. They've become fossils. Here and now they bellow and cry, swoosh and groan, or swim silently in the clear deep water. Dinosaurs pace along the shore and make the earth tremble. Birds and pterosaurs twirl about their shoulders like scattered seed.

Come along. Don't worry about the dinosaurs. Most stay off in the woods where their cries become simply a musical accompaniment to warm, peaceful evenings. But do be careful if you enter the water. Even the sharks aren't safe out there.

You'll soon see why.

The pterosaur, *Nyctosaurus*, soars in the foreground. *Nyctosaurus* is smaller than *Pteranodon*, but still stretches over 9 feet (2.7 m) between wing tips. It has a small crest and three digits on each of its wing fingers.

Look at *Pterodaustro* flying out to sea. It uses long, elastic bristles instead of teeth to filter tasty morsels from the ocean.

The Power of Flowers

Flowers entered the world of dinosaurs and redwoods like beggars at a banquet. They took root in damp, dark, and disturbed places and grew quickly. They converted shoot tip leaves into papery flags of color advertising sweet nectars and wads of delicious pollen. Birds, bees and bats came to buy with the services provided by jointed legs and sturdy wings. They carried pollen far and wide—not randomly like the wind—but drawn by the multi-colored hues of petals advertising female chalices of golden energy.

As flowers flashed their signals and speeded up the tempo of a lazy Cretaceous world, other changes followed. While some pollen grains sparked new embryonic plants to life as seeds, yet others started a frenzy of growth that encased those seeds with fleshy fruit. Once again animals came to call. They ate the fruit and left a trail of seeds to germinate where they could beneath stately mixed canopies of pines, mangroves, and broad-leaved newcomers—sometimes within the rain-filled tracks of dinosaurs.

Pre-armed to move and grow quickly, flowering plants transformed from beggars to clever thieves. When great evergreen forests flamed in a world tortured by an asteroid's fall, flowering lineages and their faithful followers stole the show.

As my favorite naturalist Loren Eiseley said, *"The weight of a petal has changed the face of the world and made it ours."*

Looking at these first flowers—these beautiful revolutionaries—I wonder what new clever creature waits in our world, ready to steal the next.

The avocado flower may resemble its distant ancestor. The barcodes of its genes tells the story.

Archaeanthus, crumpled beneath the foot of T. rex, survived to spawn magnolias and their kin.

Seashore Timeline

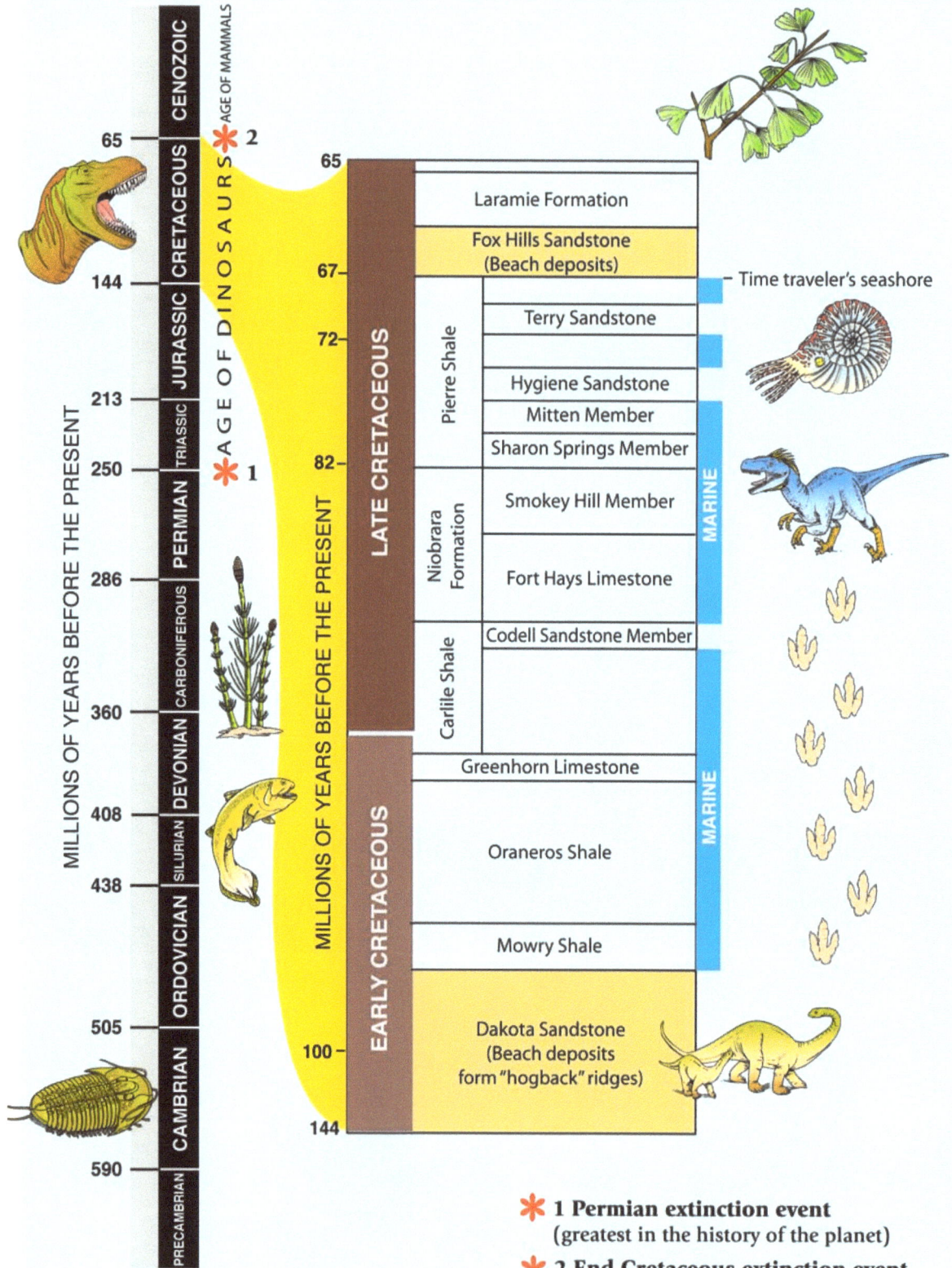

Period	Geologic Time (millions of years before the present)
CENOZOIC	65
CRETACEOUS	144
JURASSIC	213
TRIASSIC	250
PERMIAN	286
CARBONIFEROUS	360
DEVONIAN	408
SILURIAN	438
ORDOVICIAN	505
CAMBRIAN	590
PRECAMBRIAN	

MILLIONS OF YEARS BEFORE THE PRESENT

AGE OF DINOSAURS

AGE OF MAMMALS

* 2

* 1

Late Cretaceous / Early Cretaceous column

MILLIONS OF YEARS BEFORE THE PRESENT

Age	Formation	Member	MYA
LATE CRETACEOUS		Laramie Formation	65
		Fox Hills Sandstone (Beach deposits)	67
	Pierre Shale	Terry Sandstone	72
	Pierre Shale	Hygiene Sandstone	
	Pierre Shale	Mitten Member	
	Pierre Shale	Sharon Springs Member	82
	Niobrara Formation	Smokey Hill Member	
	Niobrara Formation	Fort Hays Limestone	
	Carlile Shale	Codell Sandstone Member	
EARLY CRETACEOUS		Greenhorn Limestone	
		Oraneros Shale	
		Mowry Shale	
		Dakota Sandstone (Beach deposits form "hogback" ridges)	100
			144

Time traveler's seashore

MARINE

MARINE

* 1 **Permian extinction event**
(greatest in the history of the planet)

* 2 **End Cretaceous extinction event**

Thank you for sharing this walk along the seashore. If this place haunts you as it does me, I will expect to see you again. If you find a seashell in the prairie grass think about the beautiful, savage seas in which it lived. Think, too, about how the grass on which it rests descended from those delicate revolutionaries, the flowers, whose armored and nourished seeds fell upon the world and changed it forever.   J.

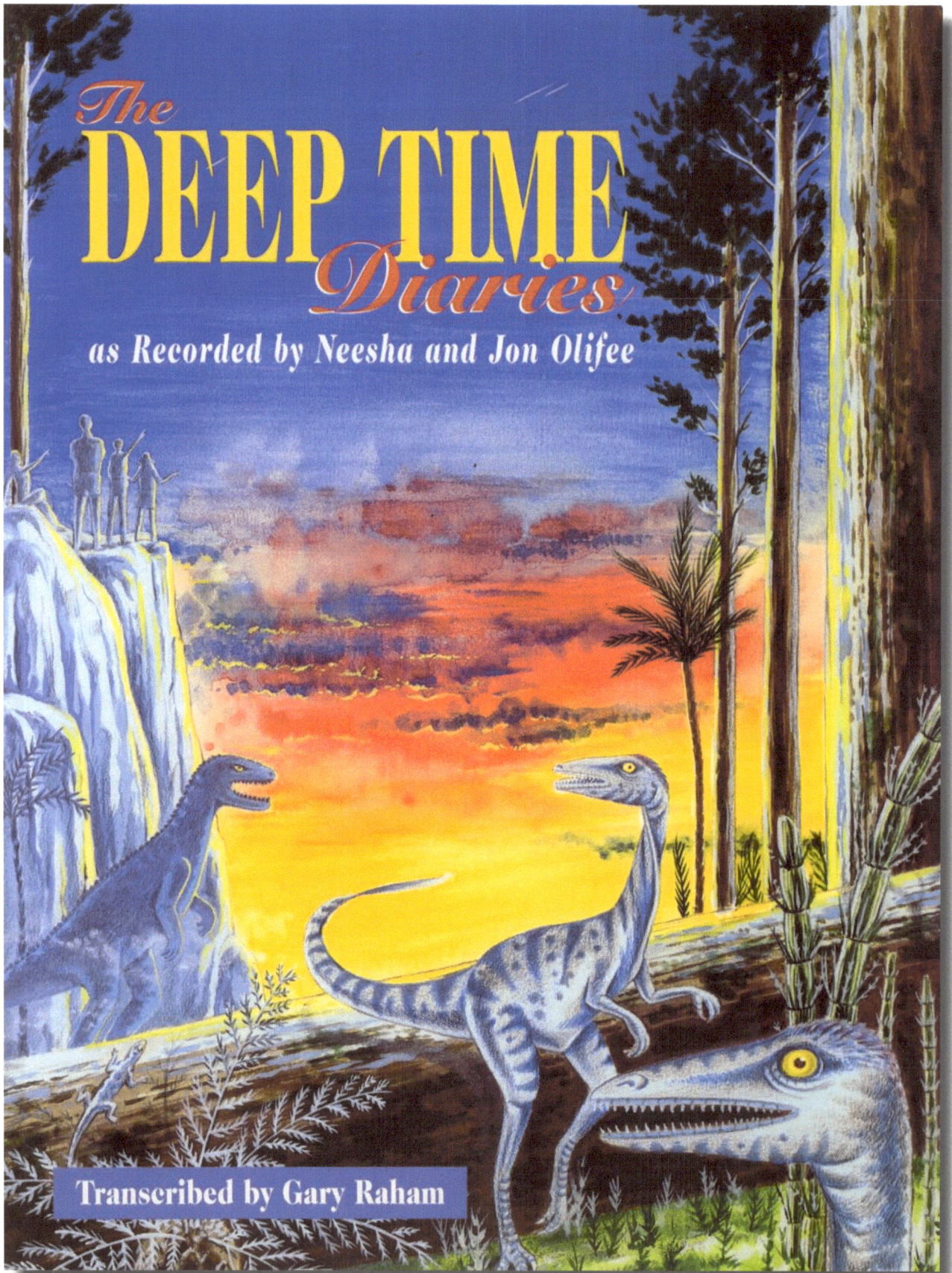

The DEEP TIME Diaries

as Recorded by Neesha and Jon Olifee

Transcribed by Gary Raham

YA NOVEL EXCERPT
The Deep Time Diaries

The Deep Time Diaries, published in 2000, *was my third book. It was fun to write and illustrate and really propelled me out of a 9-5 job and into the somewhat scary world of free-lancing—with my wife's full support, I might add. Although it didn't make me rich, the book reached a young audience for fourteen years and counting. I received gratifying feedback from kids along with immense help and support from master teacher, Vicky Jordan, who adopted this project for Wellington Junior High and other schools in the Poudre School District. Vicky and I also reached various National Science Teacher Association audiences, showing other teachers how they could tap into commercial fiction and local talent to enrich science (and English) classrooms.*

The basic plot: The book depicts the 22nd century diaries of young Neesha and Jon Olifee, two youngsters who, along with their parents, discover a time machine left behind by an alien race, the Tenoree, who visited Earth 800 million years ago. The Olifees trigger a series of time jumps that send them through deep time and into various misadventures. Along the way they—and the reader—learn lots of information about the early days of life on our planet and the fossils scattered through the corridors of deep time.

This excerpt is Jump 8 back to the Middle Cambrian, 525 million years ago. The Olifees learn about trilobites—the ancient arthropods sometimes called the "butterflies of the sea" by paleontologists. Sentient "whiz-bots" left behind by the aliens and simpler "bug-bots" assist the family on their journey.

Fulcrum publishing in Golden, Colorado, gave permission for this excerpt. Vicky Jordan and then English teacher Mark Barnes wrote "Creating Deep Time Diaries" for the April/ May 2006 issue of Science Scope magazine. As of 2014, that article is still available to NSTA members online.

Animalocaris and the Butterflies of the Sea
Jump 8: Middle Cambrian, 525 million years B.P.

Day 1: Entry by Jon Olifee

Neesha and I made the discovery of the day: We climbed up this pile of rocks at low tide looking for a little bit of shade and a spot to eat our snacks from the grease trough. Even with the oxy-booster shot[1] I was breathing hard and sweat kept trickling into my eyes. I lathered on some more sunscreen and wiped the sweat from under my hat brim. I turned to hand Neesha the lotion, but she wasn't there.

"Up here!" I heard her yell from above. When I looked up, she waved from the top of the ridge. "Stay there, grub," I said, "before you break your neck." I scrambled up a boulder pile to reach her side. She pointed out Mom and Dad a few hundred yards down the beach, but I saw something else much closer just below us. "Wow," I said. Neesha turned to look, then took a step closer to me. "What is that?" she asked.

"One monster-sized trilobite," I said. It lay in a bathtub-sized tide pool and must have been three feet long. It fluttered its legs now and then and twitched its antennae—which were half as long as its body. Its large, faceted eyes shimmered with rainbow colors in the bright sunlight.

Later, when Dad came up to draw a picture of the trilobite, we saw the feathery gills attached to each walking leg. Neesha liked its shell: lumpy, spiny, and the color of butterscotch with black and rust-colored stripes. She called him Tiger. I remember seeing fossil trilobites in school. They were always the same dull gray or brown color as the rock. Too bad colors don't fossilize because this guy would have won a prize!

By the time we left, the tide was up near the base of the rock pile. Next morning Tiger was gone, back out to sea. I hope he found his way home all right. His home is a lot closer than ours.

Day 2: Entry by Neesha Olifee

The bug-bots made some underwater gear today, so we went reef diving! We saw all kinds of trilobites—must have been Tiger's cousins. Mom called them Butterflies of the Sea.[2] They fluttered around us—in every shape, size, and color—as we swam along the reef. Some seemed to be finding worms to eat near the flowery-looking coral animals. Others filtered

"Tiger! Isn't he a beaut?" —Neesha

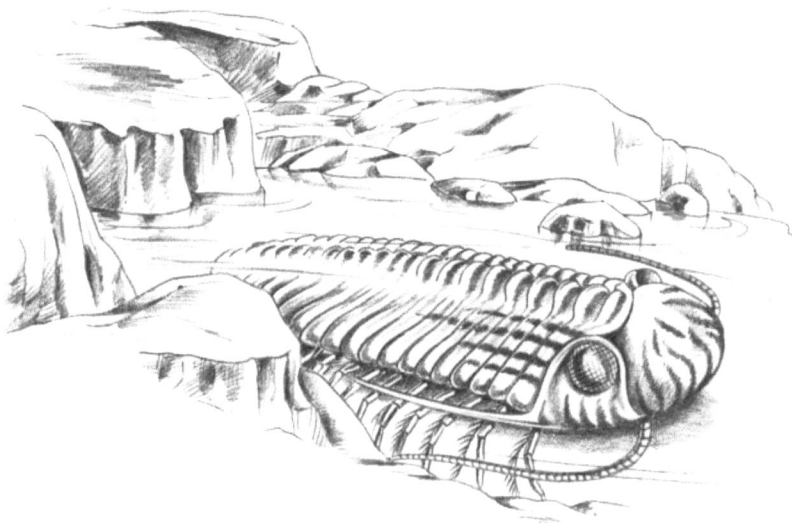

Footnote 1: *The Olifees found oxygen levels to be less than the modern norm. Thus ozone levels in the upper atmosphere were also lower.*

Footnote 2: *The paleontologist Percy Edward Raymond gave this nickname to trilobites in 1939.*

smaller goodies out of the water. Dad said trilobites have to do all the jobs fish do in the oceans of our time.

On the ocean bottom, more trilobites crawled along in the mud, some raising clouds of dirt as they tried to dig in and hide from us. It made it hard to see anything. Mom got nervous. She said if the trilobites were afraid of something our size that must mean there were some predators out there that we might want to watch out for ourselves.

Sure enough, not much later, swimming our way through a field of sea lilies[3] we noticed that most of the trilobites had disappeared. Dad motioned us to be still. Not long after, a shadow passed over us. We looked up and saw a dark, torpedo-shaped creature blocking out the light from the sun. I got goose bumps. Dad called it *Anomalacaris*, and it had a pair of tentacles hanging down from one end near a mouth that looked like a round hole ringed with a set of overlapping knives. It swam with flaps that rippled like broad, rubbery oars. It kind of coasted for a while, but suddenly made a swooping dive at the reef. A trilobite hung from underneath the shadow-beast for a moment, but pretty soon it just seemed to crumple and then disappeared into the round mouth. Only a few legs and a dark stain trailed the shadow-beast as it swam away. Yuk.

That night we made a campfire and pretended the edible white things from the grease trough were marshmallows. While Dad and Jon were trying to figure out constellations, a whiz-bot joined us and pointed in the general direction of the star Sirius. It said something like "Ten-or-ee home," then clammed up. I shivered. I remember falling asleep thinking that all those bright stars looked just like glittering trilobites dancing near a milky-colored reef.

Footnote 3: *Animals related to starfish that hooked themselves to the ocean bottom with stalks.*

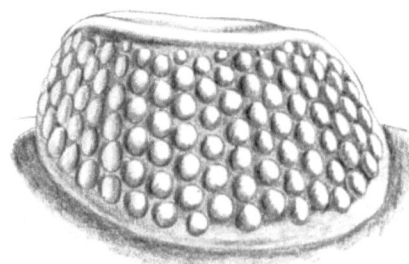
Trilobites had compound eyes long before insects did!

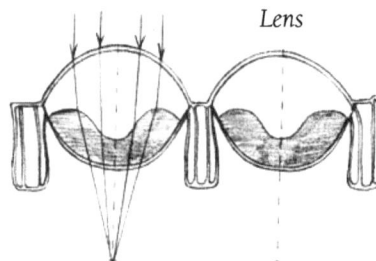
Side view of eyes

Light

Lens

Trilobite Cross Section

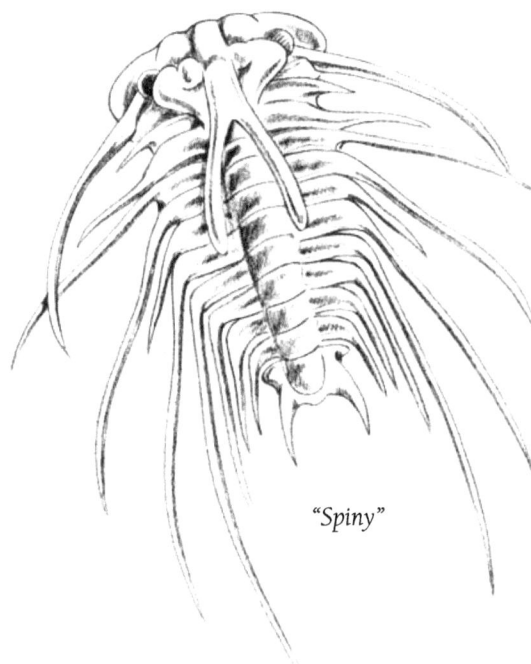
"Spiny"

For 21st Century Explorers...

Trilobites flourished for 300 million years before becoming extinct in the "Great Dying" 250 million years ago. Living relatives that share some of their features include horseshoe crabs and the tiny garden crustaceans called pill bugs, or roly-polys.

Look for pill bugs under rocks or moist leaf litter near your home. When you disturb one, it rolls up into a ball. Many trilobites could do this as well. What purpose would this behavior serve?

Flip a pill bug over onto its back. It has thirteen pairs of legs. Trilobites usually had eighteen pairs or more, but with much variation. Each trilobite leg had an attached gill for breathing underwater. Crayfish and other water crustaceans have leg gills as well. A pill bug breathes with the help of two white, bean-shaped structures on its belly, near its rear end.

Trilobites lived so long and left so many fossils throughout the Paleozoic era that they are used as index fossils. Index fossils are fossils that are only found in certain layers of rock, so wherever those layers are found, anywhere in the world scientists know they are similar in age. Find out which trilobites are index fossils for the following periods of Paleozoic time: Cambrian period, Ordovician period, Silurian period, Devonian period, Carboniferous period, and Permian period.

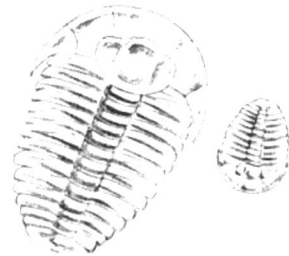

Trilobites molt to get bigger.

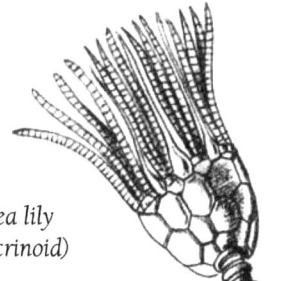

Sea lily (a crinoid)

Priapulid worm

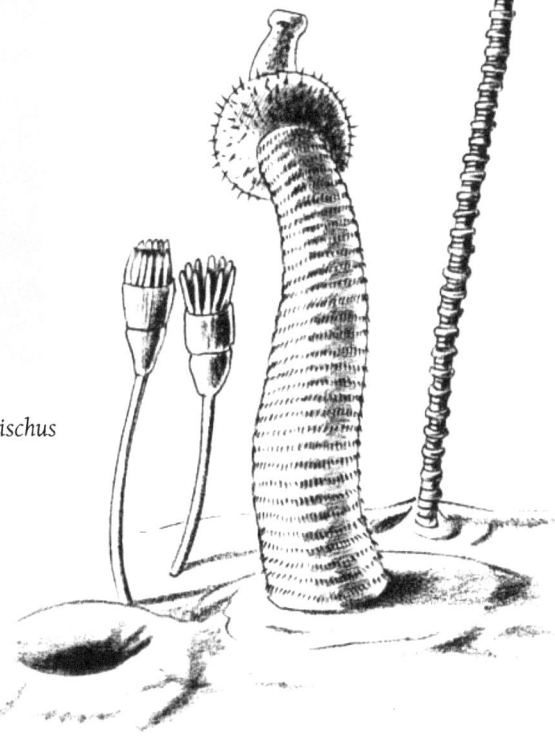

Dinomischus

Hyolithes

Rolled up trilobite
(Kainops invias)

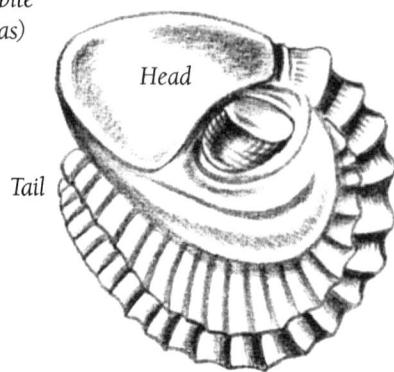

Head

Tail

Pill bug (belly up)

Breaths through
modified pleopods;
lives on land.

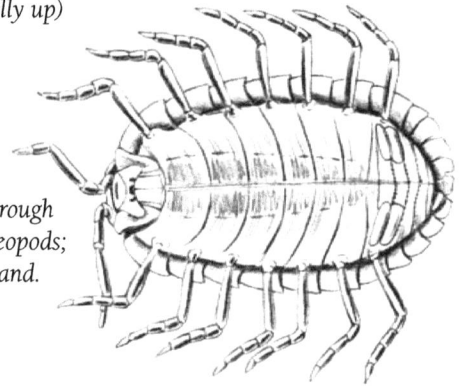

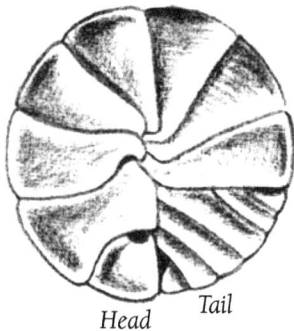

Rolled up pill bug
(Armadillidium vulgare)

Head   Tail

Trilobite
(top view)

Breathed with
gills on walking
legs; lived in the
ocean

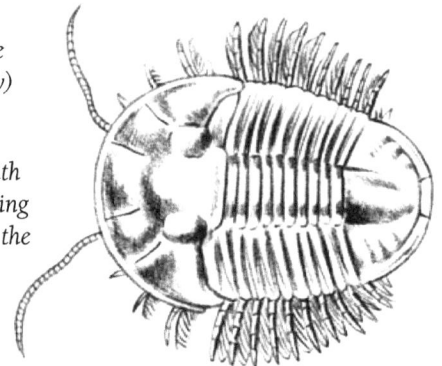

Resources and Maps

Trilobites made Earth their home for 250 million years. Not only did the dead animals become fossils, but their cast-off shells became fossils as well. As a result, you can find fossil trilobites in many places worldwide.

Always get permission from landowners before fossil hunting. Local rock clubs often lead summer field trips to nearby rock, mineral, and fossil sites. Visit a local library or museum, or ask your science teacher, to learn more about such groups.

The inset map of western Utah shows you how to get to a commercial digging quarry called U-Dig Fossils. They charge hourly or daily fees for digging fossils and provide some tools and assistance. Most of the land surrounding the quarry is owned either by the state or by the Bureau of Land Management. Check current federal and state regulations before fossil hunting on public lands.

Expect to work through hot, dry, dusty days—bring plenty of water. Nights are cool. Average elevation is 7,000 feet, and the ground tends to be rocky and gravely with scrubby plants and few trees. Roads are gravel once you leave U.S. Highway 50/6 and are in good shape unless it has rained hard recently. If you plan to camp overnight, bring your own wood for fires.

Bring rock hammers, chisels, and perhaps a pry bar. Wear a hat, and lather on the sunscreen. Sturdy boots will help protect your feet from sharp rocks and unexpected rattlesnakes.

U-Dig Fossils and the □
Trilobites of the Sevier Desert

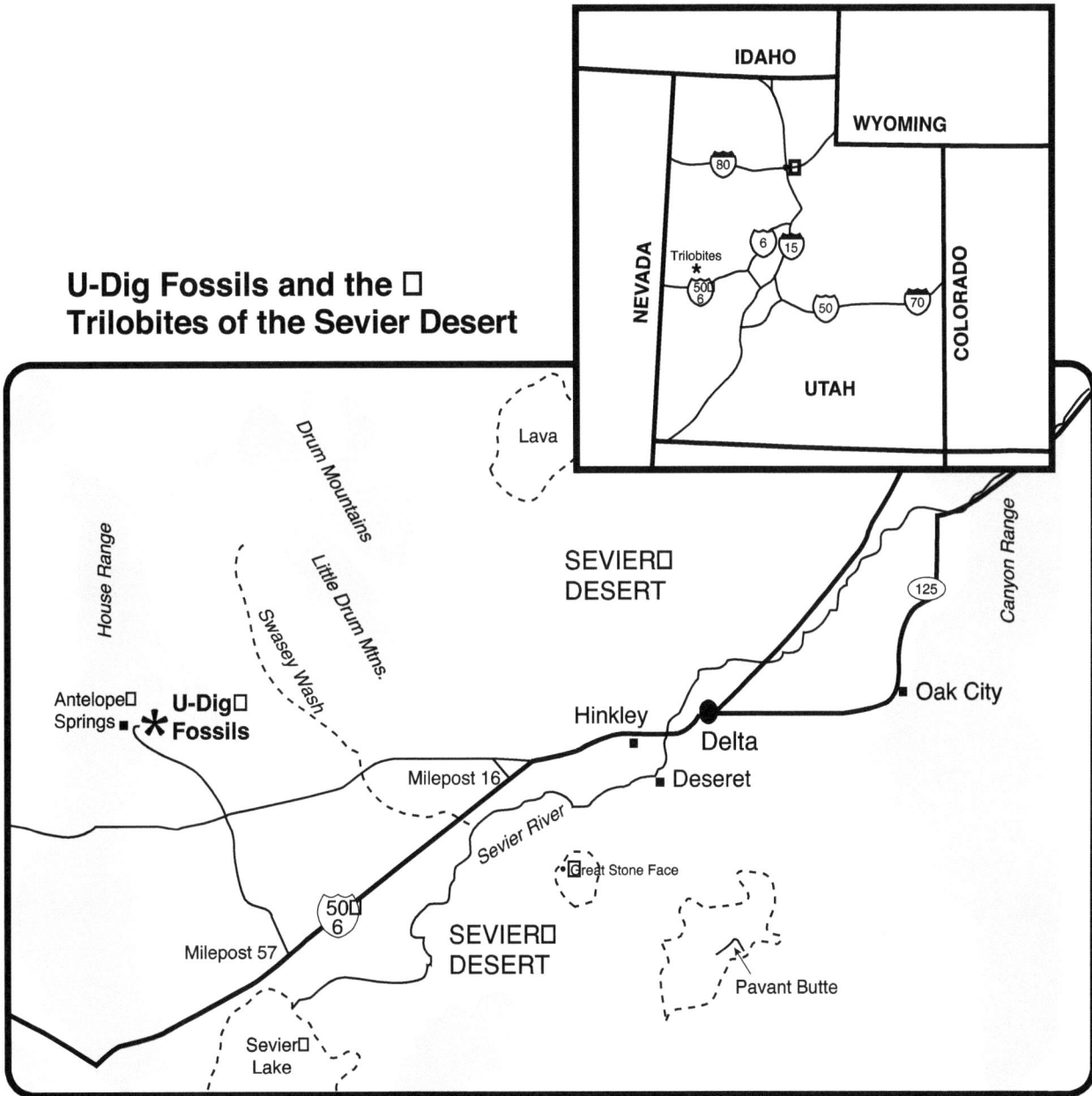

IDAHO

WYOMING

NEVADA

Trilobites
★

COLORADO

UTAH

Lava

Drum Mountains

Little Drum Mtns.

SEVIER□
DESERT

Canyon Range

House Range

Swasey Wash

125

Antelope□
Springs ■ ★ **U-Dig□**
Fossils

Hinkley

Delta

Oak City

Milepost 16

Deseret

Sevier River

Great Stone Face

50□
6

SEVIER□
DESERT

Milepost 57

Pavant Butte

Sevier□
Lake

SHORT STORY
Elvinon's Wish

I have only two formally published science fiction short stories to my name. Sometimes I regret that I didn't pursue that career path with more determination, but then again the kind of "hard" science fiction that I prefer—plausible extrapolations of life in the future based on the behavior of natural laws as we currently understand them—passed into a kind of diapause in this new century from which it's never completely awakened. Many SF authors moved on into horror and fantasy to make a living.

After twenty something years, I think Elvinon's Wish *holds up. It's built on Robert Heinlein's premise that any sufficiently advanced technology will seem like magic to the unenlightened. It's a sound premise. The story is also built on my premise that no matter how much time someone appears to have to get something done, they will invariably use all of it and then some—especially if they fall into the anal retentive state of many perfectionists.* Elvinon's Wish *also provided me with the opportunity to dally once again with my own deep time obsessions. Writers are always encouraged to "write about what they know."*

I sent this story off on a whim and it was accepted as one of the pieces in the ConAdian Souvenir book for the 52nd World Science Fiction Convention in Canada (September, 1994).

The artwork was created with acrylic and colored pencil. It was used as the back cover for a graphic short story, The Dinosaurs' Last Seashore, *but Elvinon would have been comfortable using it in his work,* Terraverte.

If you are a writer or illustrator, have fun and keep working. The end is in sight!

"Erase!" Elvinon shouted.

"Should I copy the file first?" his machine asked.

"No. Erase it. All of it."

Elvinon's labors of the last several hours blinked out of existence. In its place a field of neutral blue shimmered before his eyes. The last chords of Terreverte echoed in his ears. "I'll get you right if I have to live another seven centuries." Elvinon balled his fist and aimed it at the field terminal, but retained just enough control to realize he couldn't afford to wreck that. He kicked his worktable instead, watching with determined satisfaction as his box of record chips arced like a fountain of water and sprayed its contents in a shower of fluttering chaff around the terminal.

Elvinon limped in a small circle in front of the terminal and glared at the empty, blue workfield. "Initiate: Great Seaway, one dash A."

He appeared to be on a high, sandstone cliff overlooking an ocean that stretched to and merged with an eggshell blue sky. He breathed deeply. The view never ceased to calm him.

He waited for the pterodactyls to approach the shore, reveling in the casual grace of their glides and the deadly accuracy of their dives as they speared meals from beneath the waves. A large male settled on a sandstone outcropping to the south, folded his wings with an emperor's dignity and swung his head from side to side, shaking water from beak and bony crest in a fine spray. He screeched his desire at a passing female and the sound echoed among the rocks momentarily before being swallowed by the rhythmic murmurs of the sea.

"I need to capture more of that," Elvinon declared, in what he intended to be tones of rhetorical defiance. Instead, his voice cracked and was at least an octave too high for proper dramatic effect. He rubbed his toe, and sighed. "Perhaps tomorrow."

"Terminate," Elvinon said and the simulation vanished into the machine's magnetic-bubble reveries. In its place, Williams Lake glittered in the mid-morning sun while the familiar blue peaks of the Gore Range serrated the western horizon. Elvinon activated the shield field of his terminal to protect it from weather and animals and strode off toward the lake through reed grass and yarrow flecked here and there with the red spikes of Indian paintbrush. By the time he reached the lakeshore, his state of mind had improved greatly.

Elvinon sat down in a patch of warm sand, kicked off his shoes and buried his feet to the ankles. He looked off across the lake and concentrated on the billowy clouds above it until he could see them twist and merge in soundless white collisions. The flash of red on a blackbird's wing as it launched itself from a cattail drew his attention from sky to shore. Something glittered near twisted fingers of an aspen branch poking out of the sand. Elvinon rose to investigate.

The artifact lay mostly buried in wet sand. The portion winking to him in the bright sunlight was decorated with intricate, intertwined floral shapes. Elvinon carefully scooped sand away from the object, hoping that the rest of it was intact and of equal beauty. He was not disappointed. He soon cradled a bottle of some kind that surely deserved a spot on the art data nets.

"Ahhh…," Elvinon murmured, brushing damp sand from its surface. He polished a portion with his sleeve and held it at arm's length for critical examination. It was then, of course, that the Jinni appeared.

A diaphanous vapor twisted from the bottle, expanded, and took on form and opacity. Within seconds a young woman stood solidly on the beach. "Ahhh…," Elvinon repeated, as he examined the trim, semi-nude figure from tress-covered breasts to purple satin pants.

"Oh, thank you, Master," she said, "for releasing me from the bottle. As a reward I can grant you one wish. Your single greatest desire can be yours."

Elvinon laughed. "How did you get in that bottle? Triggered embryo development with transdimensional temporal displacement?" Elvinon scratched his chin. "No," his eyes widened, "I know! Molecular dispersion coupled with hologramic storage keyed to a tactile releaser mechanism…"

"What year is this?" she asked, frowning. Elvinon rather liked the petulant, little-girl look the frown produced.

"875 A.R.," Elvinon said.

"A.R.?"

"After Raymer." When her blank stare was followed by silence Elvinon continued. "You know…Raymer."

"Well, I don't know," she said, "but it doesn't make much difference." She sighed. "It's been a long time."

"You're trying to tell me you've been in stasis in this bottle since—before Raymer sometime?" Elvinon looked down the neck of the bottle, then at the girl. "I didn't just get sucked off the embryo trays, you know."

The girl plumped herself down cross-legged on the sand and stared out across Williams Lake. "Believe what you will," she said, "but I am grateful. You have one wish coming, if you want it."

"Look, ah…say, what is your name, anyway?"

"Call me Corlana," she said.

"Look, Corlana, how did you get in this bottle? Elvinon sat down beside her and carefully placed the bottle between them.

"Jordicon, one of my mates, is terminally jealous. At least it will be terminal when I find him." She pounded a fist into the sand, startling another blackbird from the cattails. "He put me inside."

"His sense of humor does seem a bit prehistoric."

"He has no humor and very little sense, Master…" She gave him that frown again. "What is your given name?"

"Elvinon."

"Yes, well…Master Elvinon, I do have places to go and a person to flay. Would you like that wish or not?"

"No offense, Corlana, but with my matter converter and all there's not a great deal I really need…"

"Don't underestimate me, Master Elvinon. I'm not human, you know, and I'm virtually immortal. I've learned a few tricks in the last several millennia."

"Not human?" Elvinon said. He took the opportunity to look at her carefully—and a bit wistfully—again.

"My species are great shape changers. We fine-tuned what nature gave us and can replicate nearly any life form in fine detail—as long its body mass is roughly comparable with ours. Humans are easy. I saw a big cat one time with these enormous long teeth. I've always wanted to try one of those. Do you want to see?"

Elvinon shook his head. "That's O.K. I know you're in a hurry." He coughed, stood up

and stretched, then casually put a little distance between himself and Corlana.

"Oh, don't be afraid."

"Afraid?" Elvinon laughed. "Not at all. Humans have learned a few things, too, since you've been…bottled up, I guess you'd say. The immortality thing, for example. We figured that out. Space travel. All sorts of things."

Corlana smiled politely, then stretched and shook her long hair. "It feels so good to be out." Abruptly, she rose to her feet and faced Elvinon, who was still giving her a careful examination.

"I can't believe you're not human," he mumbled.

"Well, if there's nothing you want, I'm not obligated any further…"

Elvinon's eyes widened. "There is one thing very important to me…" he absently circumnavigated a small pile of sand that he had structured with his toes. "Perhaps you could help. I'm an artist, you see, and I have this sense-o-drama thing I've been working on for—let's see now—well, it's been many decades, anyway, stretched across several centuries. I get some nice segments, you know, but I keep dithering away here and there and can't get the composition perfect." Elvinon looked into Corlana's eyes, jade green, flecked with brown, trying to see past the illusion. "How are you at artistic inspiration?"

"I've worked a lot with the e-m spectrum. UV to infrared—you name it. Why don't you show me something you've done?"

Elvinon hesitated only a moment. He hadn't had a receptive audience in quite some time. And, alien or not, Corlana reminded him just a bit of his seventeenth wife. "O.K.," he said, "my terminal is only about a mile east of here. Shall we walk?"

"I'd like that," she said, and held out her hand.

Elvinon took it automatically. If he had second thoughts, they didn't show. Corlana's hand felt quite warm and human indeed.

Elvinon hesitated to give his first command. What should I show her? He thought. A segment from the very beginning or something from "The Journey of Dinosaurs?" I wonder if aliens are as fascinated by dinosaurs as most humans are?

"What does this do?" Corlana ran her slender fingers over a segment of the terminal console. "It certainly looks impressive."

"It's a NIRS-V—A Neuromorphic-Imitating Reality Simulator. There's a model VI out now, but I like this one. It creates images and sensory experiences I conceive and broadcasts the result for others to see. Why don't you sit here," he pulled a bench from a recess in the terminal, "and you'll get the full effect." Corlana smiled and took the seat offered her. Elvinon sat in his programming chair.

"Now," Elvinon said, "I suppose I should describe a little of what this composition is about, since I'm going to show you a segment from the middle." Elvinon's eyes focused at infinity as he collected his thoughts and held his right index finger poised in the air. "Terreverte, the name of my work, literally means 'green earth.' Earth is one of those few planets blessed with conditions that allow life to flourish and I've always been fascinated by the long-term association and evolution of a planet and the living things that help to mold it." Elvinon looked at Corlana, who was stifling a yawn. "Perhaps it would be best if I just showed you, after all." He took a deep breath, wiped his sweaty hands briefly on the fabric of his body suit, commanded the NIRS-V to play…

And they were in space.

Floating. Blackness enveloped them, glittering with hard chips of starlight. Silence was broken only by the harmonies of subtle hums and buzzes that might have been the sounds of raw energy in their ears.

One chip of light grew larger. It became

a defined shape: an irregular shard of rock, miles long, trimmed with ragged mountains and pocked with empty craters whose recesses were mostly buried in ink-black shadows. The asteroid, turning slowly, its contours flickering weakly in the starlight, passed their position in space as if it had a destination and a purpose, like a shark drawn toward the blood-scent of a meal. Ahead of it lay what seemed to be a marble—a sky-blue marble, frosted with white.

The marble swelled in size until it was recognizable as planet Earth, but an Earth with the continents distorted and disturbingly out of place. South America was recognizable, but Africa lay too close to its eastern shore and North Africa was sectioned from its south end by a great channel. Antarctica was too far north and Australia too far south. North America was split east from west by a Great Seaway that lapped the feet of ragged mountains on the west and vast plains to the east.

They descended. They could hear different music now, louder and more varied. Living things were speaking to each other with threats and calls and beckonings. The many sounds distilled to one: the plaintive wails of Pterodactyls over the Great Seaway. The blues, browns and whites of Earth as seen from space transformed to a blue ocean crashing against sandstone cliffs and white-feathered dragons circling up into wispy decks of clouds. They watched the big-crested male pterosaur land on his high perch and call for his mate.

The pterosaur's call faded and transformed to the plaintive bugling of a vast herd of hadrosaurs milling near a stand of gaunt conifers, filtering the smells of the carnivores who preyed on them through their elaborate helmets of bone. In the darkening sky, a single light grew steadily larger until it began to glow like some ember blown to life…

And it continued to glow, and glowed some more…

"Ow," Elvinon's disembodied voice protested. The hadrosaurs flickered a few times, superimposed over the view of Williams Lake, then Williams Lake prevailed. Elvinon rubbed his toe. "Just a minor programming glitch," he said to Corlana, "I'll find it here in just a second."

"A very nice segment," Corlana said, "I'm impressed."

"Really?" Elvinon said, looking up from his repairs at the terminal. "You wouldn't just say that?"

"Of course not," Corlana replied, "I wouldn't say it if I didn't mean it."

Elvinon turned back to the terminal, his face frozen into an idiotic grin as he made final adjustments. "That should do it." He turned to Corlana. "Would you like to see the rest now?"

Corlana curled up in the chair and smiled. "Please. Why don't you start from the beginning?"

For the next several hours Elvinon had an admiring audience for Terreverte. The Grand Sagas of life on planet Earth unfolded with no more glitches, the Cretaceous asteroid finally struck Earth to help end the dinosaur's great reign, and the music which faded out at the end was full of hope for the future.

The late afternoon sunlight glittered in the tear on Corlana's cheek. She brushed it away, sighing.

Elvinon paced. "You know, that last part is not quite right. I must do something with it. Too melodramatic. Too…something. Then, the Cambrian section where the sea shelf breaks away and…"

"I liked it," Corlana said. "Don't fuss with it much at all."

"No." elvinon continued pacing on the little dirt path that went nowhere, except around in a circle. "Not quite right. I've got the time. I

"Pterosaur Rookery," acrylic

might as well do it right."

"I think I see what your wish is going to be," Corlana said.

Elvinon stopped in front of Corlana and looked again into her calm, green eyes. "I wish I could finish this thing." He turned and looked at the lake, an orange mirror for the setting sun. "It's an obsession, you know. Lovers come and go. I travel a lot. I have my community services, which are rewarding, of course, but I always come back to this. I have something I have to say here…"

"Oh, yes, you do," Corlana said. She uncoiled from the chair and walked over to Elvinon. She pressed her body close to his and loosely encircled his waist with her arms. "The thing you have to be sure of is that Terreverte is truly your life's work—that nothing else is more important."

Elvinon was silent only a moment. There was a sweet smell from Corlana's hair that was very distracting. He swallowed. "It is," he said. "My life's work, I mean."

"Very well," Corlana said as she drew away, "your wish is granted."

Elvinon laughed. "That was easy. If I play this composition through again will it be perfect now?"

"Don't be silly." Corlana put her hands on her hips and tossed her dark hair. "You have

Ammonite nursery, mixed media, depicting a site near Kremmling, Colorado

to create your own vision. I've simply made it possible for your wish to come true."

"And how do I know that?" Elvinon said. "Perhaps you should stay a while, Corlana, and see how I progress."

"I'll be back," she said, "after Jordicon and I work a few things out. You interest me, Elvi-non. I always have liked primate folk art. I'll be back in twenty years or so and see how you're doing."

"Twenty years will never be long enough," Elvinon sighed.

"I'm sure it will now," Corlana said, "since I've given you the gift of mortality."

"Mortality? Now wait a min-ute, Corlana…"

"Oh, no need to thank me. It was quite simple, really. When we touched I analyzed your physiology. It was a sim-ple matter to rearrange a few nucleotides here, a few histo-compatibility complexes there. And I've done my best to make the changes irreversible, so you can't be tempted."

"We need to talk about this." The veins stood out on Elvinon's neck.

Corlana's body began to fade and grow transparent. "Im-mortals make terrible artists, El-vinon. They never know when to put something down and call it finished. All you need is a firm deadline to meet." Corlana smiled, but the fading afternoon light could barely define her now. She was as insubstantial as a wraith. "Terreverte will be beautiful," she whispered, "I just know it will."

Globidens, *a shell-eating mosasur (scratchboard)*

R. GARY RAHAM

NOVEL EXCERPT
A Singular Prophecy

I began reading science fiction when I was 10 years old and have never stopped. I still have the Ace Double novel on my bookshelf: Men on the Moon *on one side;* City on the Moon *on the flip side. Copyright 1958. 35 cents.*

For me, science became the way to seek the truth: asking questions of nature with experiments, forming preliminary answers, and reveling in the new mysteries revealed. Science fiction became a way to explore what some of those mysteries might be before doing all the tedious experimentation.

I dedicated A Singular Prophecy *to my mother. She told me a fortuneteller's story about how she would not be famous, but one of her offspring would be. I used a version of that story to set my character, Ryan Thompson, on his life's path.*

This excerpt is the opening chapter. Ryan lives the dream of all paleontologists: to discover something truly amazing just by moving a little earth. Ryan gets more than he bargained for by making first contact with an alien race that first discovered Earth 72 million years ago. If you get hooked, the book is still out there in Kindle e-book land…

The mixed media illustration was used in black and white in the print book, but here it is in full color. Ryan and his alien symbiont reminisce at the end of time.

The dry earth began to swell and heave and the giant's bones quivered to life in their rocky matrix. Ryan stood, flexing his knees just enough to maintain his balance, arms folded across his chest. He glared at his friends in the quarry, sitting around like this was some kind of social. Did they expect him to do everything? "Let's get going," he said. "I don't have all night." The giant's sinuous, fish-like body rippled, her massive, skeletal head twisted free of the sediment and her right eye socket regarded him with what seemed like disdain. The giant's name was Mosie. Her jaws parted, she spoke clearly, but in an incomprehensible language. Well, isn't that just fine, Ryan thought?

His grandfather, Pops, smiled, laying down his rock hammer "Probably part of the prophecy," he said. Skeets laughed out loud, brushing a strand of hair from her eyes. She winked. "Can't you figure it out, Thompson?" Ryan felt his cheeks grow red.

Then the ground moved again--just a little undulation not far from Mosie's head--but Ryan began to shiver. He unfolded his arms. His fingers, with no arms to hold, began to fidget. The ground mounded above the new active area like some small volcano, but Ryan knew it wasn't a volcano. He wanted to close his eyes, but couldn't. His mother sat up in her freshly-dug grave. Dirt cas-

caded from her head and shoulders, balls of loam crumbled next to Mosie's animated, if petrified bones. His mother wiped some dirt from her ruined face while Ryan tried not to cry. "You know," she said, "there's something funny about Mosie's ribs."

Mosie looked down at her ribs, as if to say 'A problem with my ribs?'

"Something funny about her ribs? Mom, don't we have more important things to talk about than my fossil's ribs?" Ryan's upper lip began to quiver.

"Oh, honey," she said, "I'm sorry I had to go and leave you. Truly. But it's all right, Ryan. You'll see, it will be all right. You'll have time." She paused then, as if thinking about her words for a moment. "And even if you don't...it's still all right." His mother tried to smile, but it became nothing but a horrible, bony rictus. "You can't expect to know everything."

Ryan Thompson sat up in darkness, sweating. His heart hammered in his chest. The word "everything" echoed in his head, but the rest of the dream conversation with his mother twisted apart like smoke. He shook his head, trying to dislodge the image of his mother's half-decayed face, but instead struggled to remember her real one, before she got sick. And it had only been a year. He shivered as the sweat on his arms and back evaporated in the cool Wyoming morning.

Something had torn him from the dream. What? Pops--"Doc" to all the grad students--snored from his corner of the tent like a sleeping bear. Through the window mesh Ryan saw the Moleckson's tent sitting a hundred yards to the west in leaf-dappled moon shadow. If one of Skeets' arguments with her father had become public instead of private, he could see no evidence of it now. Even the incessant wind slept, giving the cottonwood leaves a rest. Whatever awakened him had gone.

Ryan glanced at his wrist monitor: 5:30. The sun would be up soon. His vitals' panel glowed a reassuring green. He pressed the me-

dia jack at the base of his right ear and fiddled with the scan until he found weather: Clear and hot, as usual. An early start at the quarry would be good. Pops wouldn't mind. In fact, Ryan thought, he'd probably be more surprised not to see him there early. "Mosie," his mint-new mosasaur species, would star in his first scientific paper--not bad for an in-coming UW freshman. Ryan had to get as much done as possible this field season. And he wanted to show that he knew his stuff. That he wasn't just riding on Pop's reputation.

You've got time, Ryan told himself. The bones will tell their story, one patient detail at a time. Pops had trained him well in that regard. Yet the thought that his mother hadn't had time--not nearly enough time--lingered at the edges of awareness like some sour smell.

Ryan dressed and grabbed his dusty backpack, stifling a sneeze. With his implant, he did a quick zmail check, subvocalized a short note to Dad about ordering some new software, grabbed a Nutribar for later, and sent out a web crawler to double-check a point about mosasaur limb anatomy. Then he unzipped and re-zipped the tent flap with little more than one irregular snore from Pops.

He filled his lungs with cool, sage-laced air, took a drink from his water bottle, and admired the fading stars for a few moments before padding down the familiar trail--first to the throne (a toilet seat perched on a wooden frame spanning a pit with a fantastic view of the ridge and distant hills), and then on to the ancient Cretaceous seaway-turned-to-stone and the bones of the sea beast he had found there.

Ryan cast only one glance over his shoulder on the way, blaming the morning chill for a brief flux of shivers. He found an all-tunes channel on his implant satfeed to banish the creeps and turned up the volume when Killer Moon pounded out one of his favorite songs.

By the time he got to the quarry, the sun

had bled over the horizon, casting sharp, distinct shadows. He pulled the tarp off the area he had abandoned yesterday afternoon, then folded and anchored it with a wedge of loose sandstone.

The four-foot skull grinned at him, inviting his full attention. No wonder it tended to star in his dreams. It was nearly complete, the conical jaws splayed, as if to engulf its next meal, the recurved teeth sharply defined by the morning light. Ryan wondered if Mary Anning had felt the same rush 250 years ago when her brother, Joseph, found a similar skull along the English coast near Lyme Regis. If he remembered pop's story right, she had only been eleven at the time and all fossils were unexplained "natural curiosities."

Mary would have liked Skeets. Skeets had helped him uncover some of this prize, when she wasn't with her dad--Jim--in Quarry 4. She was good at detail work and he sort of got used to having her around. She didn't prattle on all the time like some of the airhead high school kids who helped out last year. In fact, he had taken it as a personal challenge to get her to smile once in awhile. Pops said they had done a great job, that Mosie here was as big as a *Tylosaurus*, but her snout was blunter--if indeed it was a she--the teeth unique...

There was a lot to do. The sonar probes showed that there was a good chance they had 90% of a forty-foot-long animal here tucked away in sandstone. The best way to get most of it out was the same sweat-and-hammer-rocks techniques used by paleontologists for the last two hundred years, though high tech probes made it easier to find quality sites and bots made some of the grunge work easier.

His attention turned to the portion of the fossil he had detailed yesterday. Odd. Some sort of circular concretion stood in partial relief between two fragments of rib. He knew such balls of mud-turned-to-stone often contained complete fossils. Could it be something the

mosasaur had eaten? That would be a bonus. Mosasaurs were greedy enough feeders that one famous example had become enshrined in stone after choking to death on an oversized fish. Predator and prey died together. A spot on the concretion actually glinted in the morning sunlight. Had he dropped something there like a coin or piece of foil?

Ryan bent closer to get a better look. There appeared to be-- metal--where a small fragment of the concretion had chipped away: not metal ore, but processed, alloyed metal--something clearly, yet impossibly, manufactured. This concretion was no different than all the others in this outcrop. It had formed shortly after Mosie died, 70 million years ago.

Ryan's pulse raced. "Something funny about Mosie's ribs," he mumbled. He turned on the datacam, put on work gloves, and grabbed the tool belt from his pack. With the rock hammer and chisel he quickly chipped away more of the rocky nodule. A seamless metal sphere maybe six inches in diameter emerged, winking at him in the sunlight. Even though the morning chill had disappeared, he shivered. It was the same kind of shiver he felt when Pops told him about that silly prophecy eight summers ago on his first real field trip. That was when he had decided that Pops had just about the best job there could ever be.

His grandfather had poked the campfire to crackling life one cool summer night and sat down beside him on the weathered log. The firelight held the star-flecked Wyoming night at bay just beyond their shoulders. They were alone.

"Ryan, my man," he said, "You're a natural fossil hunter."

Ryan, exhausted after a hot day on treeless backcountry, remembered the warm feeling that expanded to fill his chest. "Really, Pops?"

"Hey, look at all those bones you found today. And I only had to show you once what to look for. Guess maybe that fortune teller wasn't the shyster I thought."

Ryan took the bait. "Fortune teller? What fortune teller?"

Pops stirred the fire a bit more and added another aspen log. He smiled and wrinkles fanned out from the corners of his eyes. "It was years ago. I was young, she was pretty and-- well, I kinda wanted to pull her leg a little bit and see if it would come loose. I walked in, sat down and said 'Tell me my fortune.' 'Course she tried to small-talk a story out of me, but I just handed her a five spot and said 'Tell me my future.'

"She took my hand, looked me in the eyes and smiled a little. She tried to read me like a pile of tea leaves. 'You like mysteries,' she said. 'You're good at solving them, too. Someday you'll be famous for solving very old myster-ies.'"

"Hey, that's true." Ryan's eyes grew wide. "You're a famous paleontologist."

His grandfather laughed. "I bet she knew the university had some grad students loose in the area. She saw my dusty boots and banged up nails, and had a nice vision for herself. I fig-ured she made a guess to flatter me. But then she spit on her own cake, so to speak."

Ryan smiled, waiting for Pops to finish.

"'But you know...' she said, pausing long enough for me to tilt in her direction,'...it's go-ing to be your grandson that solves the biggest mystery of all.'"

"She did? Really?" Ryan's eyebrows arched. "What'd you say?"

"I said 'and what might that be?' Well, her eyes glazed over for what seemed like a full minute, then she said, with the straightest poker face you ever saw, 'I have no idea.'"

Ryan leaned forward. "Then what?"

"Then I laughed, probably blushed a bit, mumbled something about five dollar fortunes, and left."

They both sat for awhile, quietly poking embers in the fire. Then Ryan asked: "Think you'll ever have any more grandsons?"

His grandfather's face split in a broad grin. "Better ask your folks that question, my boy."

With a smile still lingering on his face from the memory, Ryan freed a little over half of the sphere from its rocky matrix. He knew he must leave it in place--*in situ*, as Pops always said, flamboyantly emphasizing the Latin phrase, or nobody could verify--nobody would believe--where it came from. Hell, he didn't believe it and he was looking at it. But he had to touch it, without his gloves.

I created the cover image (opposite) by montaging several pictures. The background photo was taken near McCoy, Colorado, a spot I've visited with fossil groups several times in search of Paleozoic marine life. The vertebrae in the foreground are those of a mosasaur (extinct marine reptile). The reaching hand and sphere were images lifted from the web. The pair of "alien eyes" in the shadow beneath the tree were created by modifying a close-up picture of a jumping spider.

You really can't believe everything you see.

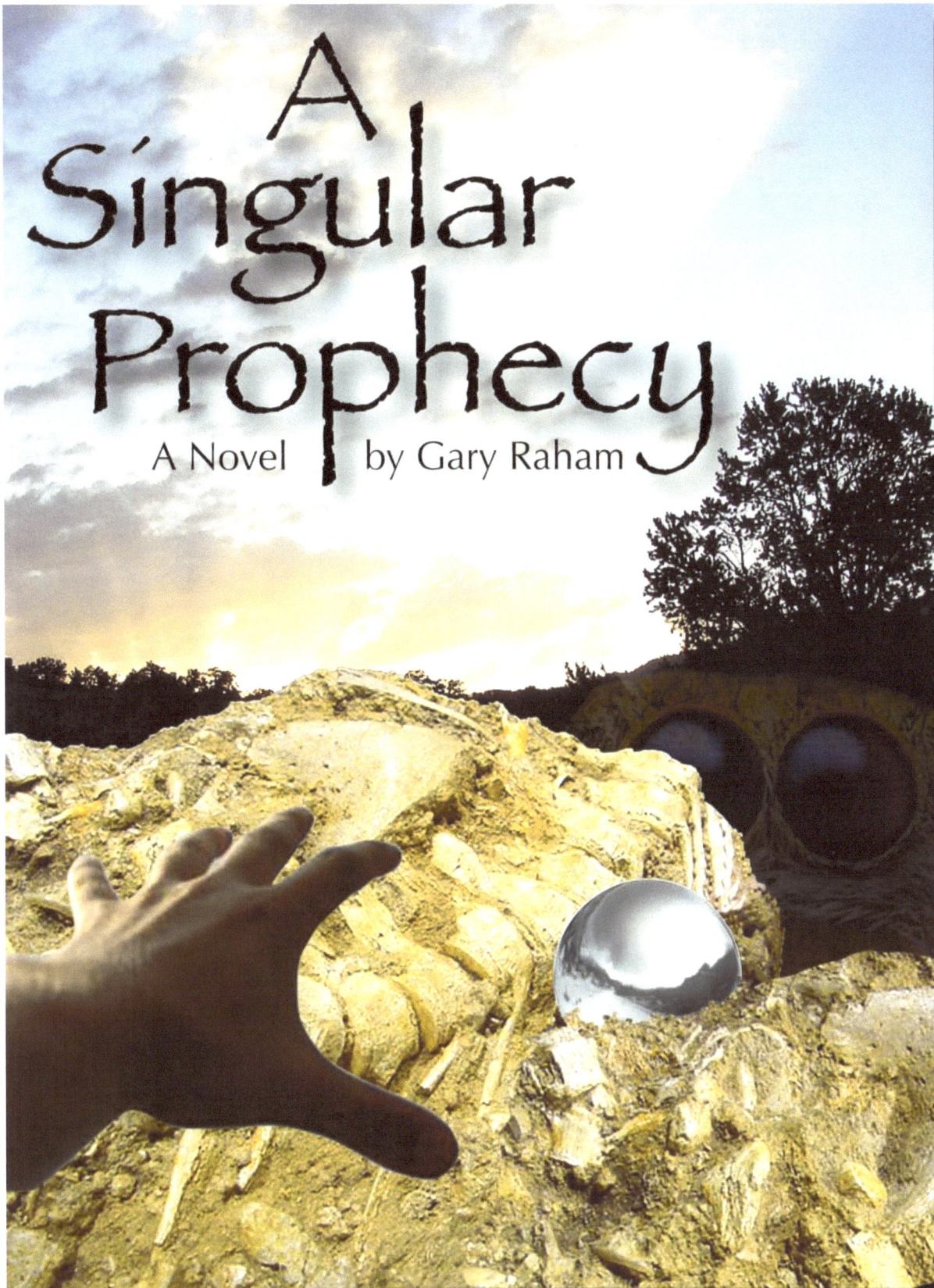

A Singular Prophecy

A Novel   by Gary Raham

Part V:
Confessing to entities & futures unknown

ESSAY
Hello out there!

I created this article specifically to precede the last essay in this book: A letter to the future. While that essay clearly addresses my compatriots on today's Earth, the following piece speaks to the real problems of either communicating with our remote descendants or making first contact with some adventurous ET. Humans have already sent some messages to the stars. Our electromagnetic "noise" began with radio and television broadcasts. In the 1970s, Carl Sagan and Frank Drake attached visual and recorded material on Pioneer 10 and 11. By the time their efforts reach an audience, our species may have vanished or morphed into other forms with different priorities.

The illustration is a photo montage of a NASA photograph of the Andromeda galaxy and a picture of the Pioneer 10 plaque. Aliens were added in Photoshop.

I confess. I'm a *Star Trek* fan. In my favorite episode, an ancient alien device penetrates the brain of Captain Jean-Luc Picard. Picard finds himself on an alien planet in the body of an individual who realizes that within little more than a generation his world and his species will die when their sun explodes. Picard lives an entire lifetime in a manner of minutes—all because another self-aware creature wanted to speak with some kindred spirit in the future and say, "I was here. I did cool stuff. Don't forget me."

Anonymous World War II soldiers scribbled, "Kilroy was here" on the walls of ruined buildings. Lord Byron carved his name into the Temple of Poseidon at Cape Sounion. Assurbanipal (685 BC – 627 BC), perhaps the first literate Assyrian king, preserved the wisdom of his culture on 30,000 clay tablets. Cave artists marked their passage through time with handprints on stone. Other than metaphorically lifting our legs in a canine-style salute to simply being here, are there other reasons to speak to the future and can we communicate successfully across vast corridors of time?

Yes and maybe.

Why communicate?

Three reasons for talking to someone or something in the future come to mind:

1.) Warn them about long-lasting dangers. Gregory Benford in his book, *Deep Time: How humanity communicates across the millennia* (1999), describes a rusted, cow-shifted plaque created in the 1960s to mark a nuclear test site in the desert. At the bottom it said: "THIS SITE WILL REMAIN DANGEROUS FOR 24,000 YEARS."

2.) Convey accumulated technical knowledge. Speaking to future generations of humans about what we know could save them a lot of time if war, climate change, or rocks falling from the sky should disrupt or destroy our civilization. The reason for conveying knowledge to future non-humans is less clear, but it could be something as simple as "life is tough. We organic creatures need to stick together against nature's implacable forces."

3.) Display our arts and smarts. As ice age tested primates we have a suite of skills and senses that make us unique. Future cultures should know the universe through our artistic prides and cultural prejudices.

How do we float a message across oceans of time, and what kind of "bottle" do we use?

Obviously, creating a plaque about nuclear dangers barely survived a half-century of wind, rain, and bovine abuse. How can we communicate with aliens or even our evolved descendants thousands or millions of years hence?

Sensory issues

As glorified, recently arboreal apes, we possess excellent color vision, probably honed by evolution so that we could deftly grab a new branch after we leaped off the old one and spot all the blushing, ripe fruit hanging from the vine. We respond to visual displays, whether they are winsome smiles across a room or pulsing graphics in a video game. Therefore, we build cities, monuments, and cathedrals and dress to impress. We send spacecraft out into the universe to snap photos of exploding stars, ringed planets, and cratered rocks. We peer at microbes and elementary particles and post embarrassing pictures on FaceBook.

To communicate to creatures of similar ilk, choosing something readily seen by keen image receptors is a good plan. However, we should realize that all creatures don't necessarily perceive our electromagnetic sweet spot. Many insects, for example, see ultraviolet light and infrared radiation—portions of the spectrum humans have to detect with other equipment or other receptors. Mollusks, like squids, respond to shifting skin colors as if they were animated billboards. And many animals play their sensory "A-game" with completely different modalities. Dogs smell. Bats echolocate. Bacteria follow magnetic fields. Whales sing. Horseshoe crabs lay eggs to the ebb and flow of moon-swelled tides.

Speed of perception may make a difference, too. Plants bend toward light, and microscopic slime mold cells sniff their kin from afar before aggregating into multicellular colonies for reproduction, but you need to measure such responses with hourglasses, not stop watches. Messages may go undetected if sender and perceiver are out of temporal synchrony.

But let's assume we want to communicate with visually oriented entities not unlike ourselves. Let's also assume they have constructed a language that can be recorded as visual symbols. Possibilities for miscommunication still abound.

Language and pictures

Contemporary humans speak somewhere between 6,000 and 7,000 distinct languages. Languages evolve and go extinct. Some experts claim a human language currently disappears with its last speaker every two weeks or so. What language should one choose for saying "Howdy" a thousand years from now?

King Ptolemy V, 2,200 years ago, hit on a great idea. He chiseled the same message in three different scripts on a 760-kilogram chunk of black granite we call the Rosetta Stone. He chose Greek, Demotic Egyptian (a kind of common man's tongue in ancient Egypt), and hieroglyphs (symbols used by priests for temples and sacred texts). Demotic Egyptian disappeared as a language in the 5th century A.D. and the secret to sacred hieroglyphs vanished as well. Greek endured, however, and served as the key to understanding the other languages. Redundancy is one key to insuring that a message survives.

Redundancy works well in the genetic code evolved by microscopic life billions of years ago. Sequences of three nucleotides along strands of DNA (and RNA) code for the amino acid building blocks of the proteins that make the enzymes and bodies we all need, but several nucleotide triplets can code for the same amino acid. One mistake in replication won't necessarily tank the development of the organism. The code may be a little garbled, but is intelligible enough to work—the solution that every message maker hopes for. A slightly imperfect protein can still make a big toe. Words missing a few vowels can still be understood. And the not-quite-perfect survivors can stumble into the future in forms that may be more suited to survival than their ancestors were. Wayward bacteria evolve into precocious primates. Classical languages morph into the King's English and rap.

Simple pictures often transcend language barriers. Note the success of travelers finding the right bathrooms because of iconic gender images and people avoiding poisons by refusing to drink products sporting skull and crossbones. When Carl Sagan and Frank Drake designed a message to aliens who might find the Pioneer 10 and Pioneer 11 spacecraft (launched in 1972 to study Jupiter and then leave the solar system), they created gold-anodized aluminum plaques with pictorial images. The images included: 1.) A representation of the most common atom in the universe: hydrogen, constructed so as to imply hydrogen's wavelength (a unit of length) and frequency (a measure of time); 2.) Pictures of a nude man and woman standing next to a

diagram of the spacecraft providing an indicator of relative size. 3.) A pattern of 15 lines radiating from a central point (meant to denote our home star). The lines come with strings of binary numbers meant to identify the periods of variable stars called pulsars, whose relative distances from the sun are indicated by the length of the radiating lines. (One assumption here is that mathematics constitutes a universal language that evolves naturally out of a basic understanding of the physical laws of the universe.) 4) A diagrammatic representation of our solar system with a line extending from the third planet from the sun to an icon of the spacecraft.

In 1977, NASA launched the Voyager spacecrafts on a related mission to study Saturn and Jupiter before leaving the solar system. Sagan helped create a somewhat different message in the form of a 12-inch gold-plated copper disk containing sounds and images meant to convey the diversity of life on Earth. Sagan and company selected 115 images, a variety of natural sounds, musical selections from various cultures and eras, and spoken greetings from humans in 55 languages. Currently, Voyager I is over 18.5 billion km from Earth (as of January 31, 2013). (See the accumulating distance at http://voyager.jpl.nasa.gov/where/index. html). Voyagers I and II will soon pierce the heliosphere—a somewhat nebulous boundary that separates outgoing solar particles (the "solar wind") from the gases of interstellar space.

The problems of cultural context

While aliens will surely differ in sensory perceptions, biology, and perhaps basic chemistry, even our future descendants will have to struggle with changes in culture: that collection of common history, assumptions, and mores that we all take as a given during our own lifetimes. Earlier in this discussion, for example, I have referred to things like *Star Trek*, World War II, Lord Byron, and rap music that

I assume my readers are aware of. Common culture provides rich sources for comparisons, not to mention humor, that enliven discourse among the enlightened, but would leave aliens and future humans scratching their heads—or other body parts.

And speaking of body parts: One thing Sagan omitted in his drawing of the naked human figures was a single line that would have depicted the vulva of his human female. He was afraid that including it might not have passed the social censorship of his time and place. He sacrificed clarity of message for being able to send any message at all.

The issues of cultural context imply that message senders must do some teaching before they can hope to communicate in order to fill in missing information and explain basic assumptions. This teaching might start by introducing numbers, equations, and other mathematical concepts before graduating to basic concepts in chemistry, physics, and biology. Finally, one could delve into language, establishing a vocabulary, dictionary, and grammatical rules (some of which may or may not be hardwired into an organism's biology).

Making the message last

Finally, it is a challenge just creating something that can endure through the vicissitudes of time. Monuments erode, metal rusts, even mountains crumble and return to the oceans. Long lasting storage devices should be: • Made of nonradioactive, stable materials • Resistant to fire • Resistant to corrosion • Insensitive to light • Impossible to counterfeit • Capable of being updated • Indifferent to technological changes • Compact and • Maintenance-free.

Engineers at the French company Arnano (based in Grenoble) believe they have a device that could last a million years. That's a good start. They first construct a thin, 20-centimeter-diameter disk made out of a form of aluminum oxide known as industrial sapphire. The disks are transparent and look like CDs (another cultural context reference). Head engineer, Alain Rey, says they are acid-resistant, scratch-resistant and more than five times stronger than tempered glass.

The message maker etches each disk with tiny images of tens of thousands of pages of information—presumably in multiple languages. The etchings are filled with platinum (more corrosion resistant than gold) and a second transparent disk of the sapphire is fused over the first. The only device the discoverer of such a disk will need is a simple magnifying lens or microscope.

Then of course comes the hard part: What do you say after that initial hello? It took Captain Picard an extra lifetime to appreciate the message he received.

How similar will other life forms be in our universe? Although we can expect great diversity, there is some reason to expect that a similar biochemistry will prevail. We are composed of the most common elements—most of them manufactured in dying stars: carbon, hydrogen, oxygen, and nitrogen—with dashes of sulfur and phosphorus. Moreover, life originated quite quickly on the early Earth: within half a billion years. Microbial life takes far longer to morph into the more complex— and for us inherently more interesting forms like mammals and magnolias.

ESSAY
A letter to the future
(To be opened in 50,000 years—give or take a millenium)

I once submitted a version of the following essay to a European editor soliciting "letters to the future" to be placed in a time capsule. I don't recall ever hearing back about which letters were chosen, so I assume mine didn't make the cut. I had fun creating the piece, however. I think, as primates who may have stumbled onto self-aware intelligence in a very haphazard fashion, we have an obligation to keep our sense of self-importance in check. Besides, it's harder to cry if you're busy laughing.

I created the illustration in 1983. I was fascinated by a quote attributed to Leonardo DaVinci: "The eye whereby the beauty of the world is reflected by beholders is of such excellence that whoso consents to its loss deprives himself of the representation of all the works of nature." I captured a tiny fragment of nature in the empty eye socket of a human skull—with a tear for the permanent loss of sight…and existence. We suffer because we can imagine all that we will never see.

Dear Entity,

Hello.

The planet on which you now stand, perch, hover, float, or paddle has made 50,000 trips around the sun since I was alive. I am a self-aware, fully organic primate approximately 99 percent chimpanzee and 1percent something else. The something else part boasts of an inflated brain and has a propensity to talk—mostly gossip about friends, relatives, rivals, and potential mates.

Fossils indicate that my species arose on a continent called Africa about 200,000 years ago— although a significant fraction of us believe that an omnipotent, primate-imitating, Intelligence pasted us together from available dust and debris about 6,000 years ago thereby instantaneously creating an awed and obedient fan base.

After a slow start as jungle tree-swingers and veggie eaters, we started making sharp rocks and added meat to our diet. We love walking and water sports and eventually tramped around the entire planet. Outside of a really bad experience with a volcano about 70,000 years ago, this has worked well for us. We are now as happy and productive as rabbits (a small, but lusty mammalian relative with large ears and a significant overbite). Seven billion of us hog approximately 50 percent of the available solar energy that plants capture from our mother star.

I think we can claim success because:

• We think about sex all the time and engage in it as much as possible.

• God loves us best—whoever She, He, or It is.

• We vigorously protect friends and family—even with our own lives—but readily kill pesky outsiders that don't look right or think wrong thoughts.

• We take twenty years to teach our kids everything we know. We live another 60 years to bitch at them when they ignore us.

• We always appreciate a good joke—perhaps because we are one.

• We have learned to be selectively rational through the practice of science—an empirical technique of asking nature questions, then running experiments to see if she wants to answer them. So far, science has allowed us to fly in objects heavier than air, make cool fireballs from broken atoms, put lots of junk into planetary orbit, read the secret code book of our genes, and discover that 97% of the universe is still a mystery.

What can you do?

We do have a few "issues" that we continue to work on:

• We are aggressive. I probably wouldn't like your looks if I saw you in person and might "punch your lights out," if I could. Although if you'd been around to beat up, we might have taken it easier on each other.

• We are picky about differences. We like our skin color, hair, clothes, political opinions, hobbies, favorite songs, Facebook pages, tweets, and gods to be just right.

• We rarely think beyond the minute at hand. We eat, drink, and make merry and let the future take care of itself—which may be why you're here and I'm not.

• We have separated ourselves from nature so that we don't get unduly scared by it. In the process, we have shoved as many species over the extinction cliff as a speeding asteroid did some 65 million years ago. Megafauna can't compete with megamalls, and we need space for roads, condos and cell towers.

We humbly refer to ourselves as *Homo sapiens* or Man, the Wise. We used to live in small groups, hunt animals and gather plants and tasty bugs. Now we prefer to live in giant cities and eat ground up cattle and deep-fried chicken bits under golden arches. We buy lots of stuff that we can't live without—such as portable picture-word processing-phone tablets, electric toothbrushes, and designer underwear.

We have learned to provide services in exchange for metal tokens and plastic cards holding records of our accumulated wealth in order to obtain the necessities described above. We call this our economy. It seems to work. Mostly. Sometimes I fantasize about hunting and gathering again, but I have become rather fond of designer underwear.

On a more personal note:

I am a male who has lived more than half of my probable life span. I have one mate, two offspring, four grandchildren, one dog, and no "love children." The latter circumstance is bad reproductive strategy for a male, nevertheless, that's the way it is. My service specialty is drawing pictures and crafting sentences designed to explain precisely what scientists know and how they work. Many people find science mystifying and prefer to believe in magic, horoscopes, and bad karma. I enjoy my work and the bounty produced by the richest country on the planet. I am basically happy and have few regrets. I do my best to ignore unpleasantness in distant places.

How about you?

Perhaps you have discovered the secret to long life or even immortality. If so, how do you fill your time? Are sex, drugs, and rock and roll sufficient? Can you interact with other sentient creatures with respect without the need to impose your will upon them? Are you organic, plastic, biomechanical, and/or biodegradable? Perhaps you are some kind of free-ranging intellect without definite form. If so, cool.

These questions are, of course, rhetorical unless you have mastered travel in the fourth dimension. If so, text me with an appropriate tachyon burst, if you have the time. It would be gratifying to know if you are a distant primate relative. That would imply our kind is more or less on the right track. But if you are 99% raven or dolphin, that's okay, too. Intelligent company is hard to come by.

I was only kidding about punching your lights out. Mostly.

Respectfully,
Gary Raham

...and that is all I have to confess...for now

THE END—of the Late Cretaceous, that is, 66 million years ago...
"Cretaceous Firestorm," (Mixed media on matteboard)

THE END—of the Paleozoic, 251 million years ago...and 95% of all life
"Permian Crisis," (Scratchboard)